The Complexity of Noise
A Philosophical Outlook on Quantum Error Correction

Synthesis Lectures on Quantum Computing

Editors
Marco Lanzagorta, *ITT Corporation*
Jeffrey Uhlmann, *University of Missouri–Columbia*

The Complexity of Noise: A Philosophical Outlook on Quantum Error Correction

Amit Hagar

ISBN: 978-3-031-01386-7   paperback
ISBN: 978-3-031-02514-3   ebook

DOI 10.1007/978-3-031-02514-3

A Publication in the Springer series
SYNTHESIS LECTURES ON QUANTUM COMPUTING

Lecture #4
Series Editors: Marco Lanzagorta, *ITT Corporation*
                Jeffrey Uhlmann, *University of Missouri–Columbia*

Series ISSN
Synthesis Lectures on Quantum Computing
Print 1945-9726   Electronic 1945-9734

The Complexity of Noise

A Philosophical Outlook on Quantum Error Correction

Amit Hagar

Department of History & Philosophy of Science, Indiana University, Bloomington

SYNTHESIS LECTURES ON QUANTUM COMPUTING #4

ABSTRACT

In quantum computing, where algorithms exist that can solve computational problems more efficiently than any known classical algorithms, the elimination of errors that result from external disturbances or from imperfect gates has become the "holy grail", and a worldwide quest for a large scale fault-tolerant, and computationally superior quantum computer is currently taking place. Optimists rely on the premise that, under a certain threshold of errors, an arbitrary long fault-tolerant quantum computation can be achieved with only moderate (i.e., at most polynomial) overhead in computational cost. Pessimists, on the other hand, object that there are in principle (as opposed to merely technological) reasons why such machines are still inexistent, and that no matter what gadgets are used, large scale quantum computers will never be computationally superior to classical ones. Lacking a complete empirical characterization of quantum noise, the debate on the physical possibility of such machines invites philosophical scrutiny. Making this debate more precise by suggesting a novel statistical mechanical perspective thereof is the goal of this project.

KEYWORDS

computational complexity, decoherence, error-correction, fault-tolerance, Landauer's Principle, Maxwell's Demon, quantum computing, statistical mechanics, thermodynamics

Contents

Preface

Quantum computers are hypothetical quantum information processing (QIP) devices that allow one to store, manipulate, and extract information while harnessing quantum physics to solve various computational problems and do so putatively more efficiently than any known classical counterpart (5). Physical objects as they are, QIP devices are subject to the laws of physics. No doubt, the application of these laws is error–free, but noise — be it external influences or hardware imprecisions — can sometimes cause a mismatch between what the QIP device is supposed to do and what it actually does. In recent years, the elimination of noise that result from external disturbances or from imperfect gates has become the "holy grail" within the quantum computing community, and a worldwide quest for a large scale, fault–tolerant, and computationally superior QIP device is currently taking place. Whether such machines are possible is an exciting open question, yet the debate on their feasibility has been so far rather ideological in character (45) (66)(110) (162). Remarkably, philosophers of science have been mostly silent about it: common wisdom has it that philosophy should not intervene in what appears to be (and is also presented as) an engineering problem, and besides, the mathematics employed in the theory of fault–tolerant quantum error correction (FTQEC henceforth) is rather daunting. It turns out, however, that behind this technical veil the central issues at the heart of the debate are worthy of philosophical analysis and, moreover, bear strong similarities to the conceptual problems that have been saturating a field quite familiar to philosophers, namely the foundations of statistical mechanics (SM henceforth). Reconstructing the debate on FTQEC with statistical mechanical analogies, this monograph aims to introduce it to readership outside the quantum computing community, and to take preliminary steps towards making it less ideological and more precise.

Amit Hagar
July 2010

Acknowledgments

The research for this monograph was supported by the NSF (Grant SES # 0847547). Any opinions, conclusions or recommendations expressed in this material are those of the author and do not necessarily reflect the views of the NSF.

Chapters 2 and 4 are based on my published articles (75) (74) that have appeared in *Philosophy of Science* and *Studies in the History and the Philosophy of Modern Physics*, respectively.

While writing these articles, as well as Chapter 3 (76) and the rest of the monograph, I enjoyed conversations with Itamar Pitowsky (HPS, HUJI), Boaz Tamir (HPS, Bar Ilan U), Meir Hemmo (Philosophy, Haifa U), Gerardo Ortiz (Physics, IUB), Emanuel Knill (NIST, Boulder) and Osvaldo Pessoa (Philosophy, USP). None of them, of course, are responsible for any mistakes or errors that might appear here, nor endorse the views expressed here. Support from Indiana University's Office of the Vice President for International Affairs (OVPIA) is also acknowledged.

This monograph is dedicated to the memory of Professor Itamar Pitowsky (1950–2010), a mentor and a friend.

Amit Hagar
July 2010

CHAPTER 1

Introduction

When quantum algorithms first appeared, it was argued that their physical realization is similar to analog computation hence unfeasible because of susceptibility to noise (102) (157). The project of quantum error correction was conceived in order to meet this challenge, claiming that quantum computers are neither digital nor analog. Rather, several error correction codes were developed that demonstrated that quantum computers are accurate continuous devices whose possible input, output, and internal states are dense countable sets (144) (149). The next obvious worry was the *cost* of error-correction: since this process is itself a form of computation, the expectation was that it would introduce additional noise to the computer. The project of fault–tolerant quantum computation meant to address this concern. Several theorems have been proven (6) (95) (96) (97) that rely on a set of requirements under which arbitrarily long computation can be fault–tolerantly executed on a quantum network if the error probability (or error rate) is below a certain threshold. Current estimations of the threshold point at an error probability of the order of 10^{-4} per computational step, which is still too low for existing technology. Given these results, the crucial open question is the following: would improvement in hardware lead to the realization of large scale, fault–tolerant, and computationally superior quantum computation, or is there still a fundamental obstacle that renders the entire project a wild goose chase?

The point is that almost three decades after their conception, and despite the aforementioned 'proofs of concept' (also known as "the threshold theorems"), experimentalists still struggle to maintain useful quantum coherence in QIP devices with no more than a handful of qubits (28). This slow progress has prompted debates on the feasibility of the entire project, yet the quantum information community has dismissed the skepticism as "ideology" (1), claiming that the problems are purely technological (89), and that their solution is contingent upon the design of quantum gates whose accuracy is below the required error threshold. From a computer science perspective, however, it is quite natural to approach the question with healthy pessimism, according to which somewhere along the way, in the scaling of the computer from several qubits (the basic information–carrying units of the quantum computer) to a large collection thereof, the computer loses its "quantumness" and with it its putative computational superiority.[1] On this view, there are *in principle* (as opposed to merely technological) reasons why, no matter what gadgets are used, large scale quantum computers will never be more efficient than their classical counterparts.

---

[1] Of course, the fact that, so far, there is no agreed–upon characterization of the physical resources behind the putative computational superiority of quantum algorithms over their classical counterparts, apart from, maybe, crude features such as 'entanglement', 'interference', or 'Hilbert space structure', to name a few, adds little to the clarity of this debate.

According to an influential view within the quantum information community (1), this pessimism is tantamount to rejecting the universal applicability of quantum mechanics. But is this influential view correct? Must quantum–computer–skeptics reject the universality of quantum mechanics? Questioning one of the premises quantum-computer–optimists rely on, namely, the famous threshold theorems, this monograph argues they need not do so: one could still doubt the feasibility of an arbitrarily large and computationally superior quantum computer while holding quantum theory intact. In fact, as we shall see, rather than quantum mechanical, the crucial obstacles are most probably thermodynamic in character.

The key insight underlying such a benign skepticism is a striking yet rarely mentioned analogy that exists between the debate on the feasibility of large scale QIP devices and the debate on the thermodynamic arrow in time in the foundations of classical SM. In the first part, the similarities between the current "active" approach to FTQEC and the radical "interventionist" solution to the puzzle of thermodynamic arrow in time in the foundations of SM will be explored. I will argue that in both approaches external noise is seen as the culprit behind the irreversible behavior of physical systems governed by reversible dynamics. On such a view, since all macroscopic systems are practically "open" and susceptible to noise, reversible behavior on macroscopic scales is unobservable unless the external noise is suppressed. In active FTQEC, noise suppression is achieved, among other things, by constantly "cooling" the computer, adding ancillas to the (open) quantum system. Remarkably, both approaches make a similar methodological choice in the way they treat noise in their models. This choice, or so I shall argue here, is what makes the famous threshold theorems physically suspect.

Once skepticism with respect to the feasibility of arbitrarily large and computationally superior quantum computers is rehabilitated, the next step is to try making it precise. Using more analogies from the foundations of classical SM, I shall argue that instead of *active* error correction, the appropriate framework for debating the feasibility of large scale, fault–tolerant, and computationally superior quantum computers should be the project of error *avoidance*: rather than trying to constantly 'cool down' the QIP device and prevent its thermalization, one should try to locate those regions in the device's state space which are thermodynamically 'abnormal', i.e., those regions in the device's state space which resist thermalization regardless of external noise.

The second part of the monograph suggests a possible skeptical conjecture and frames it in this 'passive', error avoidance, context. To be clear, while I do hint at various routes in which such a conjecture could be proven, in doing so I do not pretend to present a 'no–go' argument, nor do I spell out a quantitative skeptical attack; my sole intention is to demonstrate *why* the debate need not be deemed "ideological", and to suggest *how* it could be made precise. Moreover, even if the skeptical conjecture here suggested turned out to be true, it would not render QIP devices useless; one might still use them to outperform existing classical computers for certain designated tasks that require only few qubits. It would mean, however, that the guarded optimism that supports the feasibility of an *arbitrarily scalable* computationally superior quantum computation was premature.

In the final part of the monograph, I offer a philosophical analysis of the confusion and the double standards the quantum information community demonstrates in its treatment of error correction, namely, the way external noise was originally handled in the first threshold theorems, as well as the view of quantum error correction as a type of "refrigeration process" that eludes the constraints of thermodynamics. Underlying these cases, or so I shall argue, is a methodological predilection to regard quantum theory as universally applicable "come what may", which is also evident in the information theoretic dismissal of other approaches in discussions on the foundations of quantum theory.

CHAPTER 2

The Curse of the Open System

2.1 OPTIMISM

In 1994, quantum computing, until then a speculative research area, received a dramatic boost. Built on an earlier discovery and on an ingenious number–theoretic argument (147), a quantum algorithm was presented (143) that could factor integers into primes in polynomial number of computational steps — a sub–exponential speed–up with respect to any known classical algorithm to date.[1] The discovery drew significant attention (such an algorithm, if implemented on a large scale computer, could have dramatic consequences on current cryptography protocols) but was initially met with skepticism that portrayed the theoretician's dream as the experimentalist's nightmare (80). Roughly speaking, this skepticism was based on the indisputable fact that large scale quantum systems are noisy, and on the observation that noise, or the accumulation of errors, would impede any attempt to physically realize a (computationally superior) large scale quantum computer. In particular, it was pointed out (157) that the effects of the inevitable coupling of the putative quantum computer to external degrees of freedom put stringent constraints on the computation time, as the computer would lose its coherence (and with it lose its putative superiority) in time scales that are determined by the strength of the coupling and the state of the environment.

One may wonder why noise was seen as such a threat. After all, there is a highly successful theory of *classical* error correction that allows to protect computation against classical errors, which goes back to the early days of computer science (160).

The rub, of course, is that quantum mechanics is different than classical mechanics. In partic-ular, classical errors are discrete by their nature, boiling down to bit–flips or erasures. But a quantum state, represented, say, as a point on the Bloch sphere, is *a priori* continuous, and hence also the error is continuous. In a similar vein, quantum gates are continuous by their nature, never implemented exactly, but only up to a certain precision. Based on these differences, it was argued (102) that small errors can accumulate over time and add up to large, uncorrectable errors. On this view, quantum computation is nothing but *analog* computation, much more prone to noise, and thus actually *inferior* to digital classical computation.

Moreover, in order to protect against errors, in classical error correction, one encodes the information in a redundant way, applying a majority check on the "dressed" bits that drastically reduces the probability of error.[2] But the unitarily of quantum mechanics, i.e., the quantum no–

---

[1]The best 'sieve' algorithms for FACTORING are sub–exponential (109).

[2]The simplest classical error correction code is the *repetition* code: each bit is replaced by three of its copies ($0 \rightarrow 000$; $1 \rightarrow 111$), and a majority check can still protect against one bit–flip. The protocol fails when two or more bits are flipped. However, if the error rate for one bit–flip ϵ is assumed to be independent on each bit, then after the encoding the probability that we cannot

cloning theorem (43) (163), forbids one from copying an unknown quantum state without altering it, and it seems to prevent such a redundancy. Finally, classical error correction is based on the fact that one can acquire information on the type of error that had occurred ("the syndrome") without destroying the information encoded, but quantum mechanics seems to prevent one from doing so since measurements collapse the quantum state and may destroy the information that was previously encoded in it.

It thus came out as a complete surprise that shortly after these skeptical attacks there emerged a beautiful mathematical project (144) (149) that demonstrated the possibility of *quantum* error correction (QEC) .

2.1.1   THE BIRTH OF QUANTUM ERROR CORRECTION

The gist of QEC lies in the quantum generalization of the classical repetition code. Suppose we would like to protect a qubit against a bit–flip.[3] The protocol starts by encoding it into a larger space *without* copying or measuring it, i.e., a pair of CNOT gates and a pair of ancilla qubits are used to implement the transformation:[4]

$$\alpha|0\rangle + \beta|1\rangle \rightarrow \alpha|000\rangle + \beta|111\rangle \tag{2.3}$$

$$\tag{2.4}$$

$$
\begin{array}{l}
\alpha|0\rangle + \beta|1\rangle \\
Ancilla\ |0\rangle \\
Ancilla\ |0\rangle
\end{array}
\qquad \alpha|000\rangle + \beta|111\rangle
$$

If a bit–flip happens now, and the superposition in Eq. (2.3) becomes, say, $\alpha|100\rangle + \beta|011\rangle$, we can still extract information from the state without destroying it by measuring the *parity* of all pairs of qubits. For instance, we can measure the parity of the first two qubits with the following circuit

---

correct a bit–flip becomes $3\epsilon^2(1-\epsilon) + \epsilon^3$ (there are three possible ways to have two bit–flips and one way to have three bit–flips), and so for $\epsilon < 1/2$ we gain by encoding ((91) p. 6, (114) pp. 18–23).

[3]What follows in this subsection relies on an extremely clear exposition of FTQEC in (91). Apart from the original papers from the '90s, the interested reader can find many introductions to QEC, in varying degrees of accessibility, e.g., (61) (117) pp. 99–135 (119) pp. 425–499.

[4]The CNOT gate is a two–qubit operator where the first qubit is the control and the second qubit is the target. The action of the CNOT gate is defined by the following transformations:

$$|00\rangle \rightarrow |00\rangle \ ; \ |01\rangle \rightarrow |01\rangle \tag{2.1}$$

$$|10\rangle \rightarrow |11\rangle \ ; \ |11\rangle \rightarrow |10\rangle \tag{2.2}$$

where $|00\rangle \equiv |0\rangle|0\rangle$; $|01\rangle \equiv |0\rangle|1\rangle$, etc. The first line of the transformation signifies that when the control qubit is in the "0"–state, the target qubit does not change after the action of the CNOT gate. The second line means that if the control qubit is in the "1"–state, target qubit changes value after the action of the CNOT gate.

(adding another ancilla qubit to the code and employing two additional CNOT gates):

$$(2.5)$$

$$Code\ \{$$

$$Ancilla\ |0\rangle$$

Here each CNOT flips the ancilla qubit if the source qubit is in the state $|1\rangle$. If the first two qubits are in the state $|00\rangle$, the ancilla is left in the state $|0\rangle$. If these qubits are in the state $|11\rangle$, the ancilla is flipped twice and returns to state $|0\rangle$. Otherwise, it is flipped once by one of the CNOTs.

Note that the parity measurement does not destroy the superposition. If the first qubit is flipped, then both $|100\rangle$ and $|011\rangle$ have the same parity 1 on the first two qubits. If no qubit is flipped, the code word is still in the state of Eq. (2.3). This parity will be 0 for both $|000\rangle$ and $|111\rangle$.

Of course, a bit–flip is but *one* of a continuum of possible quantum errors that a qubit can suffer. Indeed, in general, it will undergo some unitary transformation in the *composed* system of qubit and environment that would entangle both. But it can be shown that any such unitary transformation that the composed system may undergo can be expressed as a linear combination of four basic errors known as the *Pauli group*, namely bit–flip, phase–flip, a combination thereof, and the identity. In other words, these four error types span the space of unitary matrices that can effect the qubit.[5] By performing a measurement, we collapse the combined state on one of the four 'error subspaces' hence disentangle the error from the information stored in the qubit without destroying it. This way, even though the quantum error is continuous, it will become discrete in the process of QEC.[6]

A more complicated encoding exists for phase–flip errors that uses 9 qubits and can also correct a bit–flip error and a combination of both.[7] As in the case of classical error-correction, the redundancy

---

[5] If we trace out the environment (average over its degrees of freedom), the resulting operator can be expanded in terms of the Pauli group, and we can attach a probability to each Pauli group element.

[6] In the above example, the parity measurement disentangles the code qubits from the environment and acquires information about the error. The three parities (for each qubit pair of the code word) give complete information about the location of the bit–flip error. They constitute what is called *the error syndrome measurement*. The syndrome measurement does not acquire any information about the encoded superposition, hence does not destroy it. Depending on the outcome of the syndrome measurement, we can correct the error by applying a bit–flip to the appropriate qubit.

[7] Shor's idea was to encode a qubit using nine qubits in the following way:

$$|0\rangle_{enc} = \frac{1}{\sqrt{2^3}}(|000\rangle + |111\rangle)(|000\rangle + |111\rangle)(|000\rangle + |111\rangle) \qquad (2.6)$$

$$|1\rangle_{enc} = \frac{1}{\sqrt{2^3}}(|000\rangle - |111\rangle)(|000\rangle - |111\rangle)(|000\rangle - |111\rangle) \qquad (2.7)$$

With this encoding each of the blocks of three qubits is still encoded with a repetition code, so we can still correct bit–flip errors in a fashion very similar to the above. To detect a phase–flip without measuring the information in the state we use Hadamard gates to change bases from the standard basis to the $|\pm\rangle$ basis

$$|+\rangle = \frac{1}{\sqrt{2}}(|0\rangle + |1\rangle) \ ; \ |-\rangle = \frac{1}{\sqrt{2}}(|0\rangle - |1\rangle) \qquad (2.8)$$

and measure the parity of the phases on each pair of two of the three blocks in the new basis (a phase–flip in the standard basis becomes a bit–flip in the $|\pm\rangle$ basis).

allows one to improve on the error probability for a single–qubit ϵ, since — with the discretization resulting from the error–recovery measurement — the state will be projected onto a state where no error has occurred with probability $1 - 9\epsilon$, or onto a state with a large error (single qubit, two qubit *etc.*); such a code protects against all single qubit errors. Only when two (independent) errors occur (which, in this case, happens with probability $\leq 36\epsilon^2$), the error is irrecoverable. Thus, Shor's 9 qubit QEC is advantageous whenever $\epsilon \leq 1/36$ (91).

Soon after Shor's code appeared, several alternative QECs were suggested (68) (149). It turned out that the smallest QEC that corrects a single error has five qubits, and that this is optimal (101).

Before we go on to present the active fault–tolerant version of QEC, an important remark is in place. The *crucial* assumption employed in the error models that all these codes share is that the error processes affecting different qubits are *independent* of each other. This independence assumption is also present in the treatment of errors arising from imperfect gates or from state preparation and measurements. Moreover, the first noise models that were used in the proofs of the threshold theorems (see below) introduced another independence assumption, namely an assumption about the *Markovian* character of the qubit–environment interaction. This means that the environment maintains no memory of the errors, which are thus uncorrelated in *time* as well as in qubit *location*. As we shall see, this assumption is at the heart of the debate on the feasibility of active FTQEC.

2.1.2   THE MIRACLE OF ACTIVE FAULT–TOLERANT QEC

Admittedly, error–recovery is a type of computation; hence, is not a flawless process. In the course of error correction, the application of the code itself (e.g., the parity measurements above) might suffer from imprecision. Such errors, if present, would propagate with the recovery process, leading to its inevitable corruption. Computation, moreover, requires not only storing information but also manipulating it. One could decode the (error–corrected) qubits, apply quantum gates on them, and re–encode them, but this means exposing the qubits to additional noise and decoherence. Consequently, one must perform quantum gates directly on the encoded, or "dressed", qubits.

Following specific guidelines for preventing error–propagation in the error–recovery process (i.e., verifying the ancillas, encoding the qubit, detecting the error syndrome, and recovering the error), one can demonstrate that fault–tolerant *storage* can be achieved (145) (133). With all the precautions, recovery will only fail if *two* independent errors occur in this entire procedure. The probability that this happens is still $c\epsilon^2$ for some constant c that depends on the code used.[8] Additional work has shown how to apply the gates directly to the encoded data, without introducing errors uncontrollably, and several variants of fault–tolerant universal quantum computation have been developed for different QEC codes (68).

Thus, in the few years that followed Shor's algorithm, we have seen the emergence of remarkable and beautiful mathematical results that demonstrate how to encode quantum data redundantly, how to perform fault–tolerant data recovery and how to compute fault–tolerantly on encoded states. This seems too good to be true: quantum codes exist that can correct up to t errors, where t can be

---

[8]This constant can be quite large because there are now many more gates and steps involved in maintaining the process reliable.

as large as we wish, and on which we can compute fault–tolerantly. This means that if our error rate is ϵ, then computation will only fail with probability of order ϵ^{t+1} for a t of our choice.

From the skeptical point of view that we are interested in here, however, the issue at stake is the actual *cost* of quantum error correction. If this cost scales too fast in terms of computational resources, then one has achieved nothing by employing it, as these resources would offset the putative superiority of the ideal quantum computation (without the noise). Put differently, with large number of errors t per code block, one reaches a point where the error–recovery procedure takes too much time that it becomes likely that $t + 1$ errors occur in a block, and the error correction would fail. Yet a quick calculation shows that to keep this failure probability much smaller than 1, the error rate ϵ must *decrease* with the length of the computation; the longer the computation, the more accuracy it requires.[9]

In a remarkable feat of mathematical ingenuity, this obstacle was overcome by using concatenated codes that involve recursively re-encoding already encoded bits (95).[10] More precisely, in a first–level encoding one encodes each qubit with an appropriate code. Then, for each of the codewords one encodes each of the qubits again using the same code. This process, which scales only polylogarithmically with the length of the original (ideal) algorithm,[11] reduces the effective error rate at each level, with the final accuracy being dependent on how many levels of the hierarchy are used.[12] Moreover, it was shown that an error threshold exists such that if each gate in a physical implementation of a quantum network has error less than this threshold, it is possible to perform any quantum computation with arbitrary accuracy.

Shifting the problem to the technological realm (i.e., the design of quantum gates below a certain error–threshold, which was estimated to be of the order 10^{-4}–10^{-6} errors per computational cycle), optimists could now declare:

> Therefore, noise, if it is below a certain level, is not an obstacle to unlimited resilient quantum computation ((97) p. 342).

This optimism is still present today in many expositions of FTQEC:

> The threshold theorem tells us that, in principle, we will be able to construct devices to perform arbitrarily long quantum computations using a polynomial amount of resources, so long as we can build components such that the per–gate error is below a fixed threshold.

---

[9]The number of steps required for recovery scales as a power of t, t^a with exponent $a > 1$. That means that the probability to have $t + 1$ errors before a recovery step is completed, scales as $(t^a \epsilon)^{t+1}$. This expression is minimized when $t = c\epsilon^{-(1/a)}$ for some constant c and its value is at least $p = \exp(-ca\epsilon^{-(1/a)})$. This means that our probability to fail per error correction cycle is at least p. If we have N such cycles, our total failure probability is $Np = \exp(-ca \log N\epsilon^{-(1/a)})$. For $p \ll 1$, ϵ must scale as $(1/\log N)^a$. See also (91) p. 13.

[10]This idea has, again, classical roots ((114) pp. 307–316).

[11]If the number of gates in the original algorithm is N, the fault–tolerant version thereof has $Npoly(\log N)$ gates.

[12]For example, the two–level concatenation reduces the error probability per gate p from cp^2 to $c(cp^2)^2 = c^3 p^4$ for some constant c that depends on the code, which improves the error rate exponentially as long as $p < 1/c$. If one uses k levels of concatenation, the error at the highest level is reduced to $\frac{(cp)^{2^k}}{c}$, which means the error rate decreases faster than the size of the circuit grows ((89) pp. 237–239).

> In other words, noise and imprecision of physical devices should not pose a fundamental obstacle to realizing large scale quantum computers . . . the theorem has given confidence that they can be built ((89) p. 240).

> The [threshold] theorem made it clear that no physical law stands in the way of building a quantum computer ((61) p. xvii).

Pessimists, however, were not convinced.

2.2 PESSIMISM

The threshold theorems were an important landmark in the creation of an industry as they shifted the debate on the feasibility of large scale, fault–tolerant, and computationally superior quantum computation to the technological domain.[13] But the assumptions that were crucial to the above remarkable results came under scrutiny only several years later.[14] We have already encountered one of these assumptions:

(A) *Error correlations decay exponentially in time and space.*

This assumption basically means that the environment, whose state is traced over in the computational process, maintains no memory of its states over time, and, moreover, that the error model that is assumed in the estimation of the threshold involves spatially *uncorrelated* errors. Both these features, mind you, are in stark contrast to the correlations that characterize the system whose qubits are supposed to be entangled.

Among the additional assumptions that are necessary for the validity of the threshold theorems, two are crucial to the analysis that follows:[15]

(B) *Gates can be executed in time τ_g such that $\tau_g \omega = O(\pi)$, where ω is the Bohr or the Rabi frequency.*

(C) *A constant supply of 'fresh', nearly pure, ancilla qubits is available.*

While not stated explicitly in the FTQEC literature, assumption (B) follows from the definition of a quantum gate, which is basically a unitary transformation $U = e^{iA}$ where $A = \tau_g H$ and H is the Hamiltonian generating the gate. When τ_g is scaled up, H (and hence its eigenvalues) must

---

[13]In parenthesis, it is the author's opinion that the sociological issues surrounding the rise of this industry would one day be a worthy subject for historians and sociologists of science, e.g., the fact that most of the purely mathematical results of the threshold theorems that were taken for granted, at least at the early stages, and later disseminated as "mantras", were produced by mathematicians and computer scientists; and apart from having little to do with actual physics, they were impenetrable to most working physicists.

[14]The following section relies on an analysis based on the theory of open quantum systems (14).

[15]There are other necessary assumptions, e.g., the ability to perform parallel operations or a constant error rate that shall not be discussed here.

be scaled down, and *vice versa*.[16] Assumption (C), on the other hand, *is* stated explicitly (7) and represents the ability to continuously dump the entropy produced during the noisy computational process.

The problem, however, is that there appears to be an internal inconsistency between these three assumptions: when one inquires into the derivation of the Markovian master equation that characterizes the dynamical evolution of an open quantum system (which is presumably the case at hand according to assumption (A)), one finds that the two coupling limits between the quantum computer and its environment (the singular limit and the weak limit — see (6.1)) impose constraints that are not satisfied by the other two assumptions. In particular,

- (A) and (B) are incompatible with the weak coupling limit, and so they require the singular coupling limit, which means that the reservoir (the source for the ancillas) must posses a high temperature, which then contradicts (C).

- (A) and (C) are incompatible with the singular coupling limit, and so they require the weak coupling limit, which means that the gate velocity must be slow, which then contradicts (B).

We shall now look more closely at this inconsistency claim.

2.2.1   THE HAMILTONIAN PICTURE

Active FTQEC, and, in particular, the threshold theorems proven in 1996–7, started from a phenomenological point of view, assuming a very simple, "well–behaved", error model:

> We assume that a gate's error consists of random, independent applications of products of Pauli operators with probabilities determined by the gate ((93), p. 40).

This approach was natural, given the difficulty in obtaining thresholds for models that are not phenomenological but that start, instead, from a purely microscopic, Hamiltonian, description. Such a description, however, cannot be ignored, and, as I shall argue here, may also lead to some interesting philosophical insights.

Assumption (A) arises from the upper bound on the probability p_{err} of a faulty computational path with k errors, required by the threshold theorems ((7), section 2.10):

$$p_{err} \leq c\epsilon^k(1-\epsilon)^{v-k} \tag{2.9}$$

where ϵ is the probability of a single error, c is a certain constant independent of ϵ, and v is the number of error locations in the circuit. This bound implies that the k–qubit errors should scale

---

[16]The distinction between the Bohr or the Rabi frequencies depends, respectively, on the application of constant vs. periodic fields in controlling the gate. For our purposes, however, in both cases, the crucial insight is that the gate "velocity" is directly related to the energy gap between the ground and the excited states of the Hamiltonian that generates the gate: fast gates require large energy gaps, and *vice versa*. See also (115) for a lower bound on the energy required to transform a state $|\psi\rangle$ to its orthogonal using a constant Hamiltonian and (64) for a lower bound on the amount of energy needed to carry out an elementary logical operation in a quantum computer.

as $\sim \epsilon^k$, and this can be satisfied only for reservoir correlation functions whose temporal decay is *exponential*.[17]

From a dynamical perspective, this situation can be investigated with the derivation of the quantum Markovian master equation (MME).[18] When such an investigation is carried out, the following points emerge:

- There are two types of fully rigorous derivations of quantum MME, known as the singular coupling limit (SCL) and the weak coupling limit (WCL).

- Both derivations must satisfy a thermodynamic constraint, namely that the reservoir is in a state of thermal equilibrium.[19]

- Within the SCL, this condition allows the reservoir's correlation function to be approximated by a delta function (a necessary condition for obtaining the Markovian approximation), which is the case only in the limit of infinite temperature.

- Thus, within the SCL, a Markovian approximation that satisfies the thermodynamic constraint can be justified only in the limit $T_R \to \infty$ where T_R is the reservoir's temperature.

- Physically, this means that the 'zero–memory' condition that is encapsulated in assumption (A) holds within the SCL only when the reservoir is *much* hotter than the system on the same energy scale set by the system + ancillas.

The final point, while consistent with assumption (B) (as it allows arbitrary gate velocities), appears to contradict assumption (C): where does one get 'fresh, almost pure' ancillas to dump entropy in if the reservoir's temperature is much higher than the temperature of the system? Inversely, if one requires pure ancillas, then by coupling them to the system, one must abandon the Markovian noise model in the environment.[20]

Now one may argue that the SCL, while mathematically rigorous, is nevertheless highly unphysical due to the singular character of the interaction and the infinite temperature condition. Yet the more realistic domain of WCL appears to be as unfavorable to FTQEC as the SCL:

- Within the WCL (where the reservoir's temperature is finite), one can achieve the Markovian condition in the reservoir's correlations function only after coarse graining over very long time–scales.[21]

---

[17]The localization of errors in time translates into localizations of errors in space due to the finite speed of error propagation.

[18]See Appendix 6.1.

[19]This constraint is known as the Kubo–Martin–Schwinger (KMS) condition (90).

[20]Recall that the role of fresh ancillas is to remove the entropy that is generated from the system–bath interaction. If the ancillas are already hot, this means they are initially entangled with the system, and this introduces more errors into the computational process than QEC could handle.

[21]See (41). This famous derivation of the quantum MME in the WCL is perhaps the only one consistent both mathematically and physically.

Expressed in terms of the system's gates frequency, this condition, while consistent with assumption (C), violates assumption (B), as it only allows slow, adiabatic, gates.[22]

The Markovian approximation within WCL is thus a long–time limit (at least compared to the system's gate frequency), and so one cannot expect it to hold in the time scales that are stated in the assumptions behind active FTQEC. If one wants to fault–tolerantly implement gates with a *finite* speed, one must, again, abandon the simple uncorrelated noise–model and consider non–Markovian noise.[23]

It is noteworthy that the above considerations are unavoidable if one accepts thermodynamics, as they follow from a derivation of the MME that satisfies a thermodynamic condition known as the KMS condition,[24] which is basically a restatement of the minus–first law (33), namely that a system driven out of equilibrium relaxes towards a new equilibrium state set by the external constraints.[25] Now one can, of course, doubt the validity of thermodynamics (and, in particular, the applicability of the minus–first law) in certain non–generic situations and argue that a quantum computer is such a case. Here we note that such a claim was *not* made within the FTQEC community.[26] Rather, instead of abandoning active FTQEC, a shift was made towards proving threshold theorems for non–Markovian noise (8) (151). Prior to considering this shift, let us take a short historical detour.[27]

2.3   THOSE WHO CANNOT REMEMBER THE PAST

In this section, I would like to suggest that the original threshold theorems that are based on uncorrelated noise supply yet another example of Santayana's Aphorism on Repetitive Consequences,[28] and they could have been deemed of little physical significance much earlier. Support for this claim comes from a historical analogy that exists between active FTQEC and a certain radical interpretation of a controversial school in the foundations of SM. Combined with the above results, this analogy raises doubts whether active FTQEC is the correct way to approach the realization of a large scale, fault–tolerant, and computationally superior quantum computation.

---

[22] Further analysis is possible according to the type of the driving Hamiltonian (constant or periodic) and the Bohr or the Rabi frequencies (respectively). This, however, doesn't change the requirement for adiabatic gates.

[23] Relaxing the requirement for fast gates without abandoning the Markovian condition will only generate more inconsistency: as we have seen, within the WCL, the more adiabatic the evolution, the smaller the probability for correlated noise per gate, but in active FTQEC, this probability is also inversely related to the input size. This means that if we work with adiabatic gates in active FTQEC with a Markovian noise model, we violate one of the thresholds conditions which states that the gate velocity and the input size are *independent*.

[24] See Appendix (6.1).

[25] For this reason, the common argument, according to which a quantum computer — and, in fact, *any* computer — is not in an equilibrium state, so equilibrium thermodynamics doesn't apply, is completely beside the point. That this is a non–starter is evident from the fact that the equilibrium condition is imposed only on the heat bath and *not* on the system. Moreover, using many baths with different respective temperatures is of no avail, as this doesn't resolve the inconsistency.

[26] For this and other infelicities see Chapter 4.

[27] If taken ten years earlier, this detour might have saved the quantum information industry some time.

[28] "Those who cannot remember the past are condemned to repeat it" ((139) p. 284).

2.3.1   TWO PROBLEMS IN THE FOUNDATIONS OF SM

That many things are easier to do than to reverse has been known to mankind from time immemorial. In contrast, the fundamental laws of physics are believed to indifferent to the directionality in time, allowing in principle such a reversal. One of the great challenges to mathematical physics in the 21st century is thus to construct a rigorous mathematical derivation of equations that describe irreversible behavior, e.g., Fourier's law of heat conductivity, from a classical or quantum model with a Hamiltonian microscopic dynamics (12) (30). In this respect, the fundamental question which the founding fathers of the kinetic theory struggled with in the second half of the 19th century remains open: can a reversible and deterministic dynamics fully explain the behavior of macroscopic matter? The answer is not so obvious as by now we have many examples where it is possible to derive rigorously the transport laws by adding a non–Hamiltonian component —either stochastic (42) or deterministic (62) — to the microscopic Hamiltonian dynamics, while the only Hamiltonian models for which such a rigorous proof exists are highly idealized and involve non–interacting particles (103) (104).

Another famous problem in the foundations of SM is to account for the statistical assumptions introduced into the underlying dynamics that presumably govern thermodynamic phenomena,[29] and in particular, for the assumption of equi–probability (if all states consistent with some global equilibrium constraint are equi–probable — that is, if the probability distribution in phase space is uniform over the constraint surface — then by averaging over them, one reproduces the thermodynamic relations).[30]

Roughly there are three main approaches to this problem. One can regard this assumption as *a priori* true, justifying this belief with the principle of indifference. The obvious two difficulties in this position are (1) what is the meaning of an average when one deals with an individual system, and (2) how can a subjective notion of probability affect objective notions such as entropy? One can follow Gibbs and consider an ensemble of systems, but while this approach is formally consistent (and also widely used), it doesn't resolve the first difficulty: there is usually only one actual system. Alternatively, one may follow Boltzmann's ergodic hypothesis and try to dynamically justify equi–probability, but ergodicity holds only in infinite time limits, and besides, many (thermodynamic) systems are not ergodic.[31]

An important but controversial alternative to these approaches within the foundations of SM is the interventionist school, also known as the open system approach. In its premise lies an undeniable fact: all physical system are "open", in the sense that they interact with their environment. But while this fact is acknowledged as trivial and benign by all the competing approaches to the

---

[29]That this problem is *independent* of the former sometimes goes unnoticed. See (135) for such a mistake and (73) for a criticism.

[30]There are further complications here, e.g., a distribution that is uniform with respect to one set of variables need not remain so under reparameterization, and some explanation, presumably dynamical, is required for favoring a distribution that is uniform under canonical coordinates, rather than some other distribution.

[31]Recent results in quantum thermodynamics (65) (67) (132) (150) suggest to replace the assumption of equi–probability with a simple and natural assumption about quantum entanglement.

foundations of SM, within the interventionist school, it becomes the key to the solution of the two foundational problems mentioned here.[32]

2.3.2   THE ROOTS OF THE OPEN SYSTEM APPROACH

Interventionist ideas appear already in the discussions on the plausibility of the probabilistic assumptions behind Maxwell's derivations of the velocity distribution and in Boltzmann's formulation of ergodicity. For example, Maxwell mentions "interactions with the surroundings" as a possible justification for his equi–partition theorem (quoted in (34)), and Boltzmann (29) points out that the special (highly ordered) initial conditions that lead to abnormal thermodynamic behavior, i.e., to entropy decrease, "could be destroyed at any time by an arbitrarily small change in the form of the container".

While Maxwell and Boltzmann were ambivalent with respect to the character of these external perturbations, Boltzmann's advocates in the debate on the validity of his H-Theorem that took place on the pages of *Nature* between 1894 and 1895 explicitly claimed that *random* external perturbations lead to the desired randomization of molecular motion — a randomization which was at the heart of Boltzmann's derivation of Maxwell's velocity distribution from Hamiltonian dynamics. Here is Burbury, an English barrister who took on himself the task of defending Boltzmann's views:

> Any actual material system receives disturbances from without, the effect of which coming at haphazard, is to produce the very distribution of coordinates which is required to make H diminish, so there is a general tendency for H to diminish, although it may conceivably increase in particular cases, just as in matters political change for the better is possible but the tendency is for all change to be from bad to worse (38).

Several years later Borel explained why external interventions should be instrumental in the construction of mechanical models of thermodynamic phenomena:

> The representation of gaseous matter composed of molecules with position and velocities which are rigorously determined at a given instant is therefore a pure abstract fiction; as soon as one supposes the indeterminacy of the external forces, the effect of collisions will very rapidly disperse the trajectory bundles which are supposed to be infinitely narrow, and the problem of the subsequent movement of the molecules becomes, within few seconds, very indeterminate, in the sense that an enormously large number of different possibilities are a priori equally probable (31).

According to Borel even the gravitational effects resulting from shifting a small piece of rock with a mass of one gram as distant as Sirius by a few centimeters would completely change the microscopic state of a gas in a vessel here on Earth by a factor of 10^{-100} within a fracture of a second after the retarded field of force has arrived.[33]

---

[32] For an interesting defense of a benign version of the open system approach see, e.g., (142).

[33] Such perturbations have dramatic consequences for a system whose dynamics are unstable. Michael Berry has calculated (26) that an electron in the limits of the observable universe will disturb the motion of two colliding Oxygen molecules here on Earth to

Note that Boltzmann and Borel both limit the effect of the external intervention to the instability of those initial microstates that lead to abnormal thermodynamic behavior. However, it seems unfair to bring external perturbations as *deus ex machina* to save the foundations of SM and then neglect them in the subsequent dynamics ((47) p. 231). Interventionist models — I shall call these "radical" henceforth — that have appeared in the 1950s have subsequently modified the *dynamics* of thermodynamic systems to include random external perturbations,[34] arguing that these perturbations are absolutely necessary for the construction of realistic mechanical models for thermalization:

> Statistical mechanics is not the mechanics of large, complicated systems; rather it is the mechanics of limited, not completely isolated systems (27).[35]

2.3.3   THERE IS ALWAYS A LITTLE NOISE

Those interventionists who interpret the open system approach "radically" reveal two fundamental features underlying their models. First, viewing their approach as a solution not only to the problem of equi–probability but also to the problem of irreversibility, they regard external noise as necessary for the validity of thermodynamics, and in particular for the process of thermalization; if this noise is removed, no such process can take place.[36] Second, within this radical approach there is an unexplained dichotomy between the system and its environment; while the former is treated as deterministic, the latter is allowed to include random elements.

Agreed, some interventionists may try to come up with an explanation to this dichotomy, arguing that, e.g., compared to the system, the environment is large and complex, and so its influence on the system is *effectively* random, or that the correlation established between the system and the environment during the interaction play little role in the subsequent evolution, but the crucial point here is that these explanations matter little to the conceptual problems posed in the beginning of this section (namely the problem of irreversibility and the problem of equi–probability). First, if the underlying dynamics (of the open system *and* its environment) are time–reversible–invariant, then the perturbations (or noise) can be reversed. Second, by describing these perturbations as *random* one presupposes the very randomness one would like to underpin in the first place within one's system (73).

The open system approach was rediscovered by the decoherence school in quantum theory, that lately has become the new orthodoxy.[37] There are direct historical links between interventionist

---

the extent that predictability of the motion would be lost after fifty six collisions. Note that we should not go as far as the end of the universe. Even a medium size billiard player situated one meter far from a billiard table would do the same for two elastic billiard balls after nine collisions.

[34] Famous milestones after Borel's calculation are the Spin Echo experiments (79) and the subsequent models they initiated (25) (27). This approach has recently enjoyed a revival (47) (48) (137).

[35] In that model it is nicely demonstrated how just by taking into account the walls of a container in which a gas approaches equilibrium one can achieve realistic time scales for this process.

[36] Radical interventionist models assign the environment an *active* part; as such they go beyond the usual role the environment plays in thermodynamics. For an analysis see (155).

[37] The term was coined in (35).

models and the decoherence school, but more important here is the conceptual link. In particular, the two features identified above, namely the role of noise in thermalization processes and the double standard in the characterizations of system and the environment, play a crucial part in the theory that was developed to *suppress* decoherence. This theory, as we recall, is active FTQEC.

What makes active FTQEC special — and what is emphasized in the literature on the threshold theorems — is the remarkable feat of rendering QEC, the process of overcoming the noise (by discretizing it, by disentangling it from the data through its entangling to ancillas, by dumping the ancillas, and by constantly supplying fresh ones — see section 2.1), an inexpensive process from a computationally complexity perspective. In other words, whatever cost this process may incur, the threshold theorems guarantee us that below a certain error rate, when "balancing the books" we will not transcend the putative complexity barrier between the quantum and the classical: the cost of QEC will not increase exponentially with the input size, and the quantum computer would maintain its alleged computational superiority.[38]

We can now make the analogy more explicit: think of the quantum computer as a an open system out of thermodynamic equilibrium, and on the noise as arising from the computer's interaction with a heat bath. Active QEC in this context can be seen as the attempt to "cool down" the open system, thus preventing its thermalization, and the threshold theorems are the promise that given a certain noise level, this "prevention" can be done for an *arbitrarily long* time with only moderate overhead, that is, without increasing the *overall* thermodynamic cost. Had the latter increased, we would have returned to (classical) irreversible computation, contrary to the presumably unitary (hence reversible) quantum computation that is taking place.

Now there are at least two reasons to be suspicious of this marvelous result, a result that, among other things, renders the concept of a "one-way" function (a function that is easy to compute but infeasible to reverse) obsolete.[39] The first is the latent double standard with which the system and the environment are treated here: while the interaction between the qubits + ancillas is entangling (and so non–local correlations dynamically evolve), in the interaction with the environment (or the noise model) no non–local correlations are allowed to evolve. In particular, with every computational step the environment acts as if it has "seen" the system for the first time. Note that *classical* error correction incorporates no such double standard since the dynamics that governs the bits, *as well as* the noise are *both* assumed to be yield no entanglement from the outset. Given this double standard, the original threshold theorems seem now less miraculous: if one is allowed to cheat just once in quantum mechanics, one can indeed do miracles.[40]

---

[38] Or, put differently, the protocol will be fast enough to overcome the error rate as to avoid errors–accumulation which will bring us back to a classical computational power.

[39] "One way" functions form one of the most productive discoveries in computer science. Many results in modern cryptography — especially the celebrated RSA cryptographic protocol (138) — depend on the infeasibility of recovering x from $f(x)$. In particular, while any arbitrary variable x contains as much (Kolmogorov algorithmic) information as the function $f(x)$, proving such an equivalence requires the ability to perform an exhaustive search (100) (110).

[40] Analogously, if one inserts just one *ideal* (i.e., non sensitive to thermal fluctuations) element deep in a complicated construction, one can easily use this construction to violate thermodynamics.

The second reason to be suspicious is that active FTQEC operates under the tacit assumption that external noise is responsible for the thermalization of the quantum computer:

> Quantum error correction may be thought of as a type of a refrigeration process, capable of keeping a quantum system at a constant entropy, despite the influence of noise processes which tend to change the entropy of the system ((119) p. 569).

Again, such a claim is open to the radical interpretation that it is *only* the noise that is responsible for thermalization, which means that if we eliminate the noise, no such process would take place. But surely this cannot be true. If the composed system (i.e., the quantum computer coupled to the heat bath) were left to itself, it would eventually equilibrate regardless of external perturbations, wouldn't it? Well, thermodynamics tells us that *all* physical systems out of equilibrium do so (thus obeying the minus-first law), and statistical mechanics only changes the "all" to "almost all". External perturbations (like stirring a bowl of hot soup) may accelerate this process, but, apart from radical interventionists, no one sees them as necessary for thermalization.

Note however, that if proponents of active FTQEC accepted this benign version of interventionism, then the motivation for looking at error correction as an *active* process would be somewhat blunted. In fact, if proponents of active FTQEC agreed to limit the consequence of external noise to that of perturbing the system form a state that exhibits a non–thermodynamic behavior and putting it back on a state that exhibits normal thermodynamic behavior, then the question one should ask in the context of FTQEC is not how to *eliminate* the noise, but rather how to *create* those *ab*normal states to begin with.

These considerations lead one to re–think the conceptual framework of FTQEC. If the aim is to forestall thermalization of the quantum computer, maybe active FTQEC is the wrong way to go about it. After all, one of the lessons of the foundations of SM is that if one insists on keeping the dynamics time–reversal–invariant, the reason for thermalization cannot reside in the dynamics alone and must involve the system's state space. Thus, rather than trying to actively shield the computer from external noise, perhaps one should try to create those states that resist thermalization. Such states do exist; they are allowed by theory, and (as our experience shows) in rare occasions can even be prepared in the lab.[41] The next chapter discusses this "passive" strategy and the skepticism associated with it.[42] Let us now return to the current state of affairs in active FTQEC, and to the attempts to prove threshold theorems for non–Markovian noise.

2.3.4   TOWARDS MORE REALISTIC NOISE MODELS

Combined with the attack on the internal consistency of the assumptions behind the original threshold theorems, the historical analogy proposed here suggests that the error model introduced in these

---

[41] See, for example, the spin–echo experiment (79).

[42] This passive perspective bifurcated early on from active QEC (19) (164). Whether or not this is how QEC is looked at *now* (93), is an open question. It is noteworthy that the standard formalism for QEC, known as the *stabilizer* formalism (68) provides a general framework for both *active* and *passive* QEC via its group theoretic form. But one should recall here that (1) inspite of the formal resemblance, the two strategies are physically very different, and that (2) the threshold theorems discussed here were proven solely in the context of *active* FTQEC.

theorems was inadequate. Serving its purpose in yielding provable thresholds, it nevertheless failed to characterize correctly the system–environment interaction that active FTQEC purports to describe. It is noteworthy that already in his seminal paper on classical error correction, von Neumann admitted that the statistical independence of the error per bit is "totally unrealistic" ((160) p. 90). Indeed, the exponential decay of correlations of errors seems to be an idealization of both real internal fluctuations and real random external disturbances; every real process has some intrinsic built–in delay, which means the presence of memory, or non–Markovianity.

Non–Markovianity means that instead of thinking of noise as local, i.e., as acting on the qubits *independently* of the structure of the evolution of the quantum computer (in which case this evolution only *propagates* the errors), the environment now "sees" the evolution of the computer and "learns" it. Since this evolution is non–local (recall that to protect the information from noise QEC *must* produce multi–qubit entanglement), it will eventually give rise to non–local noise (13). Although the strength of this effect can be mitigated by lowering the velocity of the quantum gates, i.e., by increasing the overall computational time,[43] its existence is a *generic* consequence of any interaction and cannot be eliminated — the unavoidable interaction with the vacuum already introduces long–range quantum memory which causes the environment to be rather malevolent by tracing the (necessarily entangled) evolution of the system — the more entangling is the evolution of the quantum computer, the more non–local is the noise.

What would be the effect of non–local noise on active FTQEC? This question is still open. First and foremost, the claim that "arbitrarily long" quantum computation can be done if the noise is below a certain threshold must be reconsidered. True, vacuum–induced decoherence is rather small in comparison to other sources of noise with which active FTQEC *can* deal, but the time scale in which the latter can now take place is certainly not "arbitrarily long", and, even more important, any statement about its length requires a delicate analysis of different time scales (i.e., the time scale in which the memory is exponentially decaying vs. the time scale in which the vacuum memory prevails).

In particular, there are two crucial consequences for the threshold estimations. First, non–Markovian noise would constrain the threshold theorems to deal with *amplitudes* and not with *probabilities* in the calculations of the error rates, as the qubits and the environment would remain entangled for longer time–scales. Since a probability is a square of an amplitude, this would have damaging effect on the error rate thresholds, as they would now become much lower than the previous (uncorrelated) case, making the entire project of active FTQEC physically unfeasible (10). One may argue that this estimation is highly pessimistic as it allows the bad fault paths to add together with a common phase and thus to interfere constructively. Most likely, or so the argument goes (92) (134), distinct fault paths would have only weakly correlated phases, and if so, then the modulus of a sum of N fault paths should grow like $\sqrt{N}$ rather than linearly in N, leading to better threshold estimations. This more optimistic hypothesis remains to be proven, yet it is hard to see

---

[43] An illustrative analogy here is a swimmer immersed in a viscous fluid; the faster her strokes, the stronger is the fluid's resistance.

how one can justify such an intricate behavior of the noise phases on an *a priori* basis, especially when one requires the exact opposite behavior for the qubits (see section 2.3.3).

Another problem that arises in threshold theorems for correlated (non–Markovian) noise (151) is that they explicitly rely on the norm of the interaction Hamiltonian $\|H_I\|$ (see Appendix (6.1)). Now, apart from the fact that this quantity is not directly measurable in experiments,[44] the requirement for low error rate means, physically, that one *assumes* that the very–high–frequency component of the noise is particularly weak, a requirement that seems not to be physically well motivated, and in some decoherence models even implies that the system and the environment are practically *de*coupled, which renders the use of QEC obsolete (11) (83) (92).

Taking stock, while the jury is still out with respect to the feasibility of active FTQEC in the presence of non–Markovian noise, one thing is certain: the initial optimism that followed the discovery of QEC and active FTQEC seems now a little premature. FTQEC is not only contingent upon technology but also dependent on the actual noise model. For some noise models FTQEC might be impossible,[45] while for other it may still be within reach for a certain amount of time. The irony is that the attempt to characterize *actual* quantum noise requires exponential resources (52). It seems that we need a quantum computer to tell us whether a large scale, fault–tolerant, and computationally superior quantum computer is possible.

2.4   OPTIMAL SKEPTICISM

Quantum information scientists would like to believe that what prevents them from building a large scale, fault–tolerant, and computationally superior quantum computer is a technological barrier. Their optimism is based on the premise that since quantum correlations are stronger than classical correlations, computationally superior quantum computers are possible *in principle*, and so it is up to us to surpass the experimental obstacles involved in controlling the noisy evolution of a large scale collection of entangled qubits. Consequently, quantum information scientists characterize pessimism about the feasibility of large scale, fault–tolerant, and computationally superior quantum computers as ideological, the claim being that such a pessimism must go hand in hand with skepticism about the universal applicability of quantum theory itself (1). Note, however, that if this were true, then the inability to create these fascinating machines would be judged rather harshly by the public.[46]

Here I have suggested that the puzzle of the feasibility of large scale, fault–tolerant, and computationally superior quantum computers resembles the puzzle of the thermodynamic arrow in time: in both domains certain states or processes that are allowed by theory remain mostly unobserved, and in both domains there exists a debate on the extent to which external perturbations are responsible for this fact.

---

[44]In fact, for otherwise reasonable noise models, the norm of the interaction Hamiltonian could be formally infinite (if, for example, the qubits couple to unbounded bath operators).
[45]For such an adversarial noise model see (85) (86) (87) (88).
[46]A similarly severe constraint on skepticism with respect to the Large Hadron Collider was criticized recently (46), the worry being that portraying the failure of the LHC as the end of theoretical physics would be damaging to the latter in case the former does fail for some reason, contingent or fundamental.

Given the above analysis of the threshold theorems and the analogy between active FTQEC and the radical interventionist school in the foundations of SM, it now appears that pessimism with respect to the feasibility of large scale, fault–tolerant, and computationally superior quantum computers is far from *ideological*. One need not abandon quantum theory in order to doubt the existence of such machines, and, on the other hand, the obstacles in realizing such machines need not be deemed purely technological. As we have seen, the feasibility of FTQEC is dependent not only on technology, but also on the actual noise, and since the distinction between the system and the noise is completely arbitrary from a fundamental perspective, active FTQEC might be the *wrong* way to approach the project. In fact, *active error correction* seems to be a misnomer: the essence of the project should be passive rather than active; errors should be avoided and not corrected.

For the purpose of advancing the debate on the feasibility of large scale, fault–tolerant, and computationally superior quantum computers, a new type of skepticism is required: one which is not too strong (as it acknowledges the universal applicability of quantum theory) and at the same time is not too weak (as it isn't contingent upon technological capabilities). Such a skepticism would be optimal as a counterexample to the received wisdom in the quantum information community. In this chapter, I have only argued that such a skepticism is *possible* and have suggested that it should be developed in the context of passive FTQEC. This task is undertaken next.

CHAPTER 3

To Balance a Pencil on Its Tip

3.1 THE PASSIVE APPROACH TO QUANTUM ERROR CORRECTION

The passive approach to quantum error correction bifurcated early on from the main path taken by quantum computer scientists (19) (111) (158) (164). The idea behind it is to look for those regions in the system's state space which are unaffected by the interaction of the system with its environment. Given the robustness of these regions, if one were able to encode information into them, this information would remain intact during the system's evolution, and one could then manipulate it, presumably using fault–tolerant gates and other techniques from the arsenal of active error correction.

Let us give a classical example to this idea using one of the simplest error avoiding codes. Assume we have an error process that with some probability flips all bits in a group, and otherwise does nothing. In this case, we can encode a classical bit as

$$0 \to 00. \quad 1 \to 01.$$

The error process will change the encoded states to

$$00 \to 11. \quad 01 \to 10.$$

Note that the parity of the two bits is conserved no matter whether the error has acted or not. So when we decode, we will associate 00 and 11 with the encoded 0 bit and 01 and 10 with the encoded 1. Note also that we will be able to decode correctly no matter how high the rate of error is! The error does not touch the invariant, parity 'space', into which we encode. That means that our encoded information has managed to completely avoid the error.

In quantum systems, conserved quantities are associated with the presence of symmetries, that is, with operators that commute with all possible errors. for one of the most common collective decoherence processes, the noise operators on n qubits are given by $S_\alpha = \sum_{i=1}^{n} \sigma_\alpha^i$, where σ_α^i is a Pauli matrix ($\alpha = \{x, y, z\}$) on the ith qubit (see Appendix (6.2)). Intuitively this means that the possible unitary errors are $\exp(it S\alpha)$. An error–free subspace exists when

$$S_\alpha|\text{codeword}\rangle = c_\alpha|\text{codeword}\rangle,$$

or, in other words, when the code space is a simultaneous eigenspace of each S_α with eigenvalue c_α. If this is the case, each unitary noise operator only introduces an unobservable phase $\exp(it c_\alpha)$ on the code space. In a trivial two–qubit example (see Appendix (6.2)), where the state of qubit 1 is stored in the state of qubit 2 without being affected by the errors, operators acting only on the second qubit commute with the error operators. In particular, if E is any one of the errors, then $E\sigma_\alpha^{(2)} = \sigma_\alpha^{(2)}E$, for $\alpha = x, y, z$. It follows that the expectations of $\sigma_\alpha^{(2)}$ are conserved. That is, if ρ is the initial state (density matrix) of the two physical qubits and ρ' is the state after the errors acted, then $tr\sigma_\alpha^{(2)}\rho' = tr\sigma_\alpha^{(2)}\rho$. Because the state of qubit 2 is completely characterized by these expectations, it follows immediately that it is unaffected by the noise.

A less trivial example is the encoding of 4 qubits ((91) p. 15):

$$|0\rangle_{\text{code}} = |s\rangle \otimes |s\rangle,$$

$$|1\rangle_{\text{code}} = \frac{1}{\sqrt{3}}(|t_+\rangle \otimes |t_-\rangle - |t_0\rangle \otimes |t_0\rangle + |t_-\rangle \otimes |t_+\rangle),$$

where $|s\rangle = \frac{|01\rangle - |10\rangle}{\sqrt{2}}$ and $|t_{\{-,0,+\}}\rangle = \{|00\rangle, \frac{|01\rangle + |10\rangle}{\sqrt{2}}, |11\rangle\}$. It is easy to see that $S_\alpha|0\rangle = S_\alpha|1\rangle = 0$ for $\alpha = \{x, y, z\}$ i.e., that the coefficients $c_\alpha = 0$). This means that both code states are invariant under collective noise. If we encode our information into the subspace spanned by $|0\rangle_{\text{code}}$ and $|1\rangle_{\text{code}}$, it will completely avoid the errors; it resides in a 'quiet' part of the system's space, dubbed as the decoherence–free subspace.

When do these 'quiet' subspaces exist and how can they be constructed? The trivial two–qubit example suggests a general strategy for finding a noiseless qubit: first determine the commutant of the errors, which is the set of operators that commute with all errors. Then find a subset of the commutant that is algebraically equivalent to the operators characterizing a qubit. The equivalence can be formulated as a one–to–one map f from qubit operators to operators in the commutant. For the range of f to be algebraically equivalent, f must be linear and satisfy $f(A^\dagger) = f(A)^\dagger$ and $f(AB) = f(A)f(B)$. Once such an equivalence is found, a fundamental theorem from the representation theory of finite dimensional operator algebras implies that a subsystem identification for a noiseless qubit exists (98).

A noiseless subspace is thus perfect for maintaining quantum information. Yet one may argue that operations such as manipulating this information, or even storing it initially, require access to that subspace, hence disturb it. In order to eliminate such inevitable disturbances, the general strategy is to keep the information stored in the QIP device in its 'quiet' subspace using dynamical methods. The idea here is a generalization of the famous 'quantum Zeno effect' (118): one aims to keep the system in a subspace by repeatedly projecting the system into it. This projection has a nonzero failure probability so that the cumulative probability of repeated successful projection may be expected to fall with the number of projections. The quantum Zeno effect provides a means of maintaining the cumulative probability of successful projections arbitrarily close to unity. The basic principle is illustrated in the following simplified example ((19) p. 1543): consider a quantum system initially in state $|0\rangle$ which rotates into $|1\rangle$ with angular frequency ω. The state at time t (in the absence of any projections) is $\cos \omega t|0\rangle + \sin \omega t|1\rangle$. If we project this state into the subspace spanned by $|0\rangle$, then the probability of successful projection is $\cos^2 \omega t$. If we project repeatedly n times between $t = 0$ and $t = 1$, i.e., at time intervals $\delta t = 1/n$, then the probability that all projections will be successful is

$$\left(\cos^2\frac{\omega}{n}\right)^n \approx \left(1 - \frac{\omega^2}{n^2}\right)^n \rightarrow 1 \quad \text{as} \quad n \rightarrow \infty \tag{3.1}$$

Thus, if the projections are performed with sufficient frequency, then the state may be confined to the decoherence–free subspace with arbitrarily high probability. In quantum mechanics, projection operations correspond to measurements on the system so the above may be loosely phrased as "a frequently observed state never evolves" or "a watched pot never boils". A similar analysis holds for any unitary evolution of a state initially lying in the subspace.

In recent years, it has been shown (158) that a variety of techniques in quantum control theory could be harnessed to keep the quantum information 'at bay', i.e., prevent it from straying outside the boundaries of the decoherence–free subspace. The open question, of course, is the complexity cost of these techniques, i.e., how do the resources they require scale with the size of the system, and it is here, or so I shall argue, where a skeptical conjecture on the feasibility of large scale, fault–tolerant, and computationally superior quantum computers can be formulated and be made precise.

In order to motivate this conjecture, let's make another short detour into the foundations of classical SM.

3.2   LESSONS FROM THE FOUNDATIONS OF CLASSICAL STATISTICAL MECHANICS

Classical SM is the branch of theoretical physics that aims to account for the thermal behavior of macroscopic bodies in terms of a classical mechanical model of their microscopic constituents with the help of probabilistic assumptions. Chapter 2 has pointed out several analogies that exist between the foundations of this branch and the debate on the feasibility of large scale, fault–tolerant, and computationally superior quantum computers. In this chapter, we shall rely on the foundations of classical SM again, and, in particular, on the combinatorial approach of Boltzmann, for the purpose of motivating the skeptical conjecture in the context of the passive approach to quantum error avoidance.

In his response to criticism mounted against his (in)famous attempt to dynamically 'derive' the second law of thermodynamics (also known as the H–theorem), and, in particular, to Loschmidt's reversibility objection, Boltzmann shifted to what is known today as the probabilistic, or combinatorial, approach ((156) pp. 55–64). In this approach, instead of trying to underpin the thermodynamic arrow in time in the system's dynamics, as he did with his H–theorem, Boltzmann attempted to re–state the empirical fact encapsulated in the second law of thermodynamics by introducing several assumptions with which he described the approach to equilibrium as a tendency for the system to evolve towards ever more probable macrostates, until, in equilibrium, it has reached the most probable state.

The details of Boltzmann's combinatorial approach as they are spelled out in his papers and summarized succinctly in (156) need not concern us, and it is certainly not my intention to enter into a debate about its relative plausibility viz. other approaches to the foundations of classical SM.[1] The important lesson I am interested to highlight, instead, is that with the help of several assumptions — e.g., a distinction between macro– and micro– states, the equiprobability of each microstate in a given energy shell, a specific 'carving' of μ space into finite cells, or partitions,[2] the limited applicability of these assumptions to ideal gases — Boltzmann succeeds in attributing probabilities to volumes of phase space, and, consequently, he re–describes thermalization in statistical mechanical terms.

In Boltzmann's combinatorial approach, the tendency to evolve from improbable to more probable states is presented as a fact of experience rather than the consequence of any theorem: *abnormal* thermal behavior (e.g., fluctuations out of equlibrium) is quite possible and consistent with the dynamical laws but also highly improbable. In other words, in Boltzmann's combinatorial approach *abnormal* thermal behavior is deemed *rare* where 'rare' is judged relative to the standard Lebesgue measure.

Yet whatever Boltzmann had in mind as a complete solution to the puzzle of the thermo-dynamic arrow in time, his combinatorial approach falls short of producing it: clearly, questions about any dynamical evolution (thermodynamically *abnormal* or otherwise) must be answered with appeal to the system's initial state and Hamiltonian; arguments from measure theory alone are thus insufficient (156).

Among the efforts to complete Boltzmann's attempt, perhaps the most well–known is the one supplied by the Ehrenfests, who suggest that Boltzmann somehow relied on the ergodic hypothesis in his reply to Loschmidt.[3] But the same ergodic hypothesis would imply that the system cannot stay inside the equilibrium state forever and thus there would necessarily be fluctuations in and out of equilibrium. Consequently, one would have to state that the tendency to evolve from improbable to probable states is itself a probabilistic affair; something that holds true for most of the initial states, for most of the time, or as some or other form of average behavior.

The Ehrenfests failed to supply such a statistical extension of Boltzmann's H–theorem. A famous landmark towards that end was Lanford's theorem (103),[4] but more recently, a group of beautiful mathematical results, known as the fluctuation theorems (53) (54), have brought us as close as one can get to the fulfillment of Boltzmann's combinatorial program. These results give an analytic expression for the probability, in a nonequilibrium system of finite size observed for a finite time, of a thermodynamic abnormal behavior in the reverse direction to that required by the second law of thermodynamics.

More precisely, for a dissipation function that represents a generalized entropy production, and it is defined as:

$$\overline{\Omega}_t(\Gamma)t \equiv \ln\left(\frac{f(\Gamma, 0)}{f(\Gamma(t), 0)}\right) + \int_0^t ds \frac{\frac{dQ}{dt}(s)}{K_B T_{res}} \tag{3.2}$$

---

[1]On this issue, there already exists a vast literature (58).

[2]Clearly, Boltzmann's combinatorial approach only makes sense with finite partitioning of μ space.

[3]If the ergodic hypothesis holds, a state will spend time in the various regions of the energy shell in phase space in proportion to their volume.

[4]Lanford's theorem cannot be regarded as the final step in the development of the combinatorial approach since it holds only for a ridiculously short time and only in the highly unphysical Boltzmann–Grad limit of a rarified hard–sphere gas.

where $f(\Gamma, 0)$ is the initial phase space distribution function, K_B is Boltzmann's constant, and dQ/dt is the rate of heat gained or lost per unit time by the system from a thermostat,[5] the theorems relate the probabilities of observing time averaged values of $\overline{\Omega}_t$, for a period of time, t, equal to an arbitrary value A, relative to $-A$:

$$\frac{Pr(\overline{\Omega}_t = A)}{Pr(\overline{\Omega}_t = -A)} = e^{At} \tag{3.3}$$

This ratio is exponential in the length of the averaging time t and the number of degrees of freedom in the system.[6]

The result is exact for classical systems and quantum analogues are known. It confirms that for large systems, or for systems observed for long times, the second law is likely to be satisfied with overwhelming (exponential) likelihood: positive entropy production is overwhelmingly more probable than a negative one.[7] Note that this also implies that as physical devices are made smaller and smaller the probability that they will run thermodynamically in reverse to what one would expect, increases exponentially with decreasing system size and observation time.

It is tempting to apply the combinatorial approach, and, in particular, the lesson of the fluctuation theorems to the domain of (reversible) quantum computing:[8] the decoherence–free subspace is regarded as 'an island of stability' in which reversible (unitary) dynamics can go on without disturbance, resisting thermalization. In this sense, this 'island' harbors abnormal thermodynamic evolution, and it is natural to expect it to decrease in size as the system becomes macroscopic. Using this analogy, one might put forward the following skeptical conjecture:

C_0 *Decoherence–free subspaces supporting quantum states that allow computational superiority become exponentially rare as the system's size increase.*

But this is all too quick, for the combinatorial approach is still plagued with two persistent problems.

First, while the fluctuation theorems allow one to describe abnormal thermodynamic behavior as *rare* (mind you, in complete agreement with our experience), one should still specify 'rare' according to what measure, and subsequently justify such a choice of measure. Let's call this problem *the measure problem*.

Second, and more importantly, one must also provide a link from the alleged scarcity of this much–sought–for behavior to the difficulty in observing or controlling it: on the uniform measure

---

[5]The thermostat is viewed as being much larger than the system of interest and can therefore be regarded as being in thermodynamic equilibrium at a temperature T_{res}.

[6]Recall that entropy production is extensive.

[7]It should be emphasized that this theorems 'derive' macroscopic asymmetry from microscopic symmetry only by sneaking in an asymmetric assumption (as Boltzmann did in his H–theorem). Contrary to Boltzmann's early slumber, however, the authors of this theorem are very explicit about this ((54) p. 1580). Also, these results are given relative to the SRB measure, which is an extension of the micro–canonical measure to systems far from equilibrium. As such, they fall prey to the well known problem of the justification of measure that plagues the foundations of classical SM, and is discussed below.

[8]Two such combinatorial arguments recently appeared in the literature. In the first (129), a uniform measure on all possible experiments is assumed, and it is shown that the norm of a random entanglement witness decreases exponentially with the size of a quantum system. In the second (69), it is argued that *most* (where 'most' is interpreted again relative to a specific measure) many–body entangled quantum states are useless for computational speed–up.

in phase space, for example, the vast majority of states are equilibrium states. A steam engine requires a heat source and sink of different temperatures. One could point out that states that have such temperature differentials are rare, according to this measure, but this says nothing about the feasibility of constructing a steam engine. Similarly, if we take the Haar measure, then, for a large environment, by far the vast majority of system–environment states are ones in which the system is highly entangled with its environment. Nevertheless, we believe that we can readily prepare systems in states that are pure, or close to pure.

The difficulty in providing such a link between the abstract representation of states of a physical system and the creation or manipulation of such states by the scientist signifies the limitations of the applicability of the statistical mechanical formalism to the observer. That such a link is missing from the foundations of classical SM should be regarded as one of the key problems in that field, yet so far it has been rarely mentioned in the literature.[9] Because of its conceptual similarity to a well–known problem in quantum theory, and for lack of a better name, I dub it here as *the SM measurement problem*.

Within the foundations of classical SM these two problems are still open. With respect to the first, I have no pretension to solve it here. Let me just mention that attempts to justify the choice of measure that rely on (classical) dynamical considerations are, on final account, plainly circular, stated as they are as valid for *almost all* cases, where the term "almost all" is defined again by an appeal to the very measure one would like to justify in the first place. Given this circularity, one may as well abandon the attempts to find such an *a priori* justification, and instead admit that the relation between probability in classical SM and the standard Lebesgue measure is purely an empirical matter. What I would like to suggest here instead is a solution to the second problem that not only makes the first less urgent but also allows one to draw the analogy between the combinatorial approach to the foundations of SM and the debate on the feasibility of large scale, fault–tolerant, and computationally superior quantum computers.

3.3   TO BALANCE A PENCIL ON ITS TIP

Clearly, $\mathbf{C}_0$ is still open to objections on the basis of the two problems mentioned above, i.e., the measure problem and the SM measurement problem: it doesn't specify the measure relative to which the decoherence–free subspaces are deemed rare, and it doesn't tell us anything about the computational complexity resources we need to invest in order to create them or maintain a state therein. In what follows, I propose the following refinement that will allow us to replace statements about 'rarity' (with respect to a given measure) with statements about computational complexity. In doing so, I believe one could sidestep the measure problem — dropping altogether the term 'rare' — and at the same time provide a link between the 'size' of the decoherence–free subspace and the difficulty associated in maintaining a state in it.

---

[9]One area in the foundations of classical SM where this problem has received some attention is the debate on Maxwell's demon, and in particular, the question whether the demon itself is subject to the laws of thermodynamics and statistical mechanics.

First, we need to address the notion of a 'creation of a state'. Recall that while in classical SM a state of a physical system is defined as a probability distribution on phase space, in quantum mechanics physical states are defined as subspaces in the Hilbert space. Combined with the observation that a quantum computation is nothing more than the standard Schrödinger evolution in the computational basis from the state $\Psi_0 = |0\rangle_1 |0\rangle_2 |0\rangle_3 \dots |0\rangle_L$ to the desired output state, this conceptual difference prompts one to replace the statistical mechanical probabilistic notions with a new operational, information–theoretic, notion of 'hardness', explicating the latter in terms of computational complexity resources. Since the basic building block in this context is the number of computational steps, given by the number of 1–2 qubits gates that are required to generate an arbitrary state from Ψ_0, the key question is how does this number scale with the size of the input (the number of qubits L). Following the familiar definitions from computational complexity theory, we can call states whose generation requires only polynomial resources *easy*, and states whose generation requires exponential resources *hard*.

In parenthesis, it is important to be reminded that not every easy state would serve for computationally superior quantum computation. To achieve the latter, we need specific entangled states that can make the probability of retrieving the result of the computation better than mere guessing (78) (128). The best example we have today for such a state is the following:

$$\frac{1}{\sqrt{2^L}} \sum_{c=0}^{2^L-1} e^{\left(\frac{2\pi iac}{2^L}\right)} |c_{L-1}\rangle |c_{L-2}\rangle \dots |c_0\rangle \tag{3.4}$$

where L is a natural number and $0 \le a \le 2^L - 1$. This superposition is the one that allows Shor's algorithm to factor numbers into primes more efficiently than any known classical counterpart. It is a sum, with equal weights, over all possible 0, 1 states of a given length. In its form, this superposition is similar to many such superpositions of equi–weighted sums. What makes all the difference here, however, are the phase factors $e^{\left(\frac{2\pi iac}{2^L}\right)}$ (128).

Apart from the number–theoretic argument behind Shor's algorithm, its quantum mechanical significance lies in the fact that the state (3.4) can be generated with a number of 1–2 qubit gates polynomial in the input size L. And yet, at least in the network model of quantum computation, where arbitrary entangled states can generated from the input state Ψ_0, there is no *general* way of doing so in a number of computational steps (i.e., 1–2 qubit gates), which is bounded by a polynomial in L.

After quantifying how hard it is to create a certain quantum state in an ideal situation, we need to consider next the non–ideal situation, where noise is allowed. In such a case, the theory of quantum error avoidance tells us that if the quantum state resides in its decoherence–free subspace, it will remain intact throughout the dynamical evolution. Thus, we need a way to quantify again how hard it is maintain a computationally superior easy state inside the decoherence–free subspace.

To this end, we may return to the aforementioned method of the quantum Zeno effect that allows one to repeatedly project the quantum state back to the noiseless subspace. Recall that for a given input size, this method succeeds in stabilizing the state, increasing the probability of the successful projections with the frequency of the projections. Now, what happens if we keep the success probability (or the projection frequency) fixed but increase the input size? How does the projection frequency (or the success probability) scale in this case?

Since there exists a trade–off between the frequency of the projection that keeps a noise–resilient state in the appropriate decoherence–free subspace and the noise–level (or fault–tolerance) one aims for, to achieve better fault–tolerance one must increase the projection frequency. Consequently, if one succeeded in demonstrating exponential scaling of this frequency with the input size for a computationally superior state, it would be tantamount to demonstrating that as the dimension of the system (i.e., the input) increases; the noise–resilient subspace becomes exponentially hard to maintain. More precisely, one needs to show that for a fixed success probability, the projection frequency increases exponentially with the size of the input (or, inversely, for a fixed projection frequency, the success probability decreases exponentially).[10] Such a proof would also entail that the unfavorable scaling of these resources offsets the alleged computational superiority that the specific state may allow in an ideal situation without the noise.[11]

Another possible way to quantify how hard it is to create a decoherence–free subspace stems from the algebraic formulation of the noiseless subsystem definition. This formulation allows one, under certain conditions, to reduce the problem of finding whether certain a subsystem is noise–resilient to other linear algebra problems. Progress in this context would amount to reducing the problem of finding a decoherence–free subspace to a computational problem whose complexity is known. So far, this route has shown only limited success. For example, the problem of deciding whether a subsystem is noise–resilient with respect to initialization procedures was shown (94) to be polynomially equivalent to the problem of finding a matrix with orthonormal columns in a linear space of matrices. The complexity class of the latter, however, is still unknown.

What I suggest, thus, is an analogy between the combinatorial approach and quantum error avoidance that would (1) replace statements about rarity with a complexity–theoretic notion of hardness, (2) provide an operational meaning to the 'size' of the decoherence–free subspace, and (3) allow us to interpret this 'size' in terms of the resources required to keep the computationally superior state therein.

Using such an operational meaning, one can offer a slightly modified skeptical conjecture:

C_1 *The physical computational complexity of keeping a computationally superior quantum state in its noise–resilient (decoherence–free) subspace increases exponentially with the size of the input and offsets the putative computational superiority such a state might afford.*

---

[10]The projection frequency is ultimately related to energy (64) (113).

[11]Calculating the resources for the Zeno effect may be too stringent because one may also look at the error density in the code space and actively correct small errors *before* applying the full Zeno measurements, so that the frequency of these becomes now lower and scales constantly with the code size. In such a case, one would have to estimate a threshold for the noise density so that if it is smaller than this threshold, one can reduce the frequency of the Zeno measurements and save on resources. Here again, though, the discussion on the kind of noise one introduces (Markovian or non–Markovian) becomes important.

One possible objection is that as a conjecture based on complexity considerations, C_1 might as well be applied against *classical* computers, but since these are evidently scalable *in principle*, the entire line of argument presented here is unwarranted. Note, however, that C_1 is directed solely at the marginal, putative *superiority*, of QIP devices *viz. a viz.* their classical counterparts. In other words, C_1, if true, would make large scale QIP devices effectively *classical*, hence cannot (and should not) be applied against the scalability of classical computers.[12]

The advantage of C_1 over C_0 is clear, but one may still object that, lacking a full–fledged theory of physical computational complexity, the suggested link between computational complexity resources and the putative difficulty in maintaining a computationally superior easy quantum state inside its noiseless subspace is not exhaustive. In other words, since there seems to be no way to uniquely formalize experimental procedures or physical interactions into quantum unitary evolutions and classify them further into computational complexity classes, the map, if there is such, between the former and the latter is clearly not bijective; hence, C_1 lacks rigor, and as such, it is no better than C_0.

In response, let me just say that while a complete theory of physical computational complexity is indeed lacking, we can still formalize *some* dynamical evolutions and physical interactions in the language of quantum theory and classify these further into different complexity classes. For example, a 2–body, nearest neighbor interaction is known to be easy (in the computational complexity sense) while some many–body interactions that cannot be broken efficiently into the former are known to be hard (161). Acknowledging as I am that the field of physical computational complexity is still in its infancy, my only aim here is to initiate a line of thought whose final goal would be a new computational complexity outlook on physics that would hopefully yield a better understanding of the relationship between the two disciplines. In this sense, a conjecture such as C_1 provokes one to study particular examples, using them a starting point for a dialectic between theoreticians and experimentalists, wherein the former present the latter with the challenge of creating increasingly complex quantum states.

Faced with this conjecture, theoreticians and experimentalists would have to re–examine known quantum systems with a new goal in mind, namely, that of quantifying the complexity resources of their respective states. Once such evidence would accumulate, and in order to lend support to C_1, quantum computing skeptics could look for a trade–off between those quantum states that can be prepared efficiently in the presence of noise (but have no computational advantage) and those that cannot (but do have such an advantage). Only then could the question of the feasibility of large scale, fault–tolerant, and computationally superior quantum computers be said to be well–posed.

3.4   PRACTICAL VS. PHYSICAL POSSIBILITY

This chapter was intended to demonstrate how the debate on the feasibility of large scale, fault–tolerant, and computationally superior quantum computers could be made more precise by taking into consideration some lessons from the foundations of classical SM. Ignoring these lessons, many

---

[12]The scalability of the latter is ultimately bounded by contingent physical laws and boundary conditions.

quantum information scientists believe that the failure to realize a large scale QIP device would indicate new physics:

> If it is possible to continue scaling up such devices to a large size, the issue of the absence of cat states becomes more pressing ... If generating such states is successful, then the existence of, in essence, Schrödinger's cats will have been shown. Such states are, however, more sensitive to the effects of phase decoherence, but this seems to be a technical, not a fundamental, problem. Therefore, if it becomes impossible to make such states or to build a large scale quantum computer for non–technical reasons, this failure might indicate some new physics ((28) p. 1014).

C_1, however, offers a different outlook, and if true, would prove both parties in the current debate wrong: contra the pessimists, large scale, fault–tolerant, and computationally superior quantum computers would be deemed physically possible, in the sense that they are consistent with the laws of physics, in particular with the laws of quantum theory and statistical mechanics, yet, contra the optimists, for the very same reason one rarely encounters abnormal thermodynamic processes in the macroscopic regime, large scale *and* computationally superior quantum computers would be rendered impossible in practice: since preparation and manipulation of those states that can allow computationally superior and fault–tolerant quantum computation is a computationally complex task itself, QIP devices could not be scaled arbitrarily; as their size increased, they would lose their putative computational power and become no more efficient than their classical counterparts.

The possible failure of scaling up QIP devices need not render them useless. The traditional notion of efficiency (based on the distinction between polynomial and exponential growth) is an asymptotic notion referring to computations on unboundedly large inputs, and thus may not be appropriate in assessing the feasibility of particular computations in practice. For example, with more than 40 noiseless qubits, a QIP device would presumably allow simulations of quantum systems beyond what is possible with classical physics. FACTORING might require much more resources, but if a quantum computer could factorize a 1000–digit integer in a reasonable time, it may still exceed the abilities of any classical computer for the foreseeable future albeit that the factorization of 2000–digit integers might be infeasible on any computer, quantum or classical.

This means that one should be careful when comparing quantum computation to classical computation. Conjectures such as C_1 are concerned with asymptotic limits, and not with arbitrary cut–offs that are contingent on human interests or commercial targets. Therefore, even if C_1 turned out to be true, there might be some computational tasks for which a QIP device with a small number of noiseless qubits would still outperform a classical counterpart of a similar (or of any) size. No matter how outstanding, however, such a performance would still fall short of realizing the unqualified hope, shared by many quantum information scientists, for *unlimited* resilient quantum computation.

In closing, a final word of caution: so far, the number of noiseless qubits we have managed to produce in the lab is between 2 and 8 (159) (28).[13] This tremendous achievement is clearly consistent

---

[13]That almost a decade has passed from the former to the latter demonstrates the difficulty involved in such a project.

with *any* statement whatsoever one could make today on the feasibility of large scale, fault–tolerant, and computationally superior QIP devices.

CHAPTER 4

Universality at All Cost

4.1 OVERVIEW

Quantum information science has enjoyed an exponential growth both in funding and in worldwide exposure in the last three decades. During that time, the paradoxes of quantum non–locality and the measurement problem that have been pervading the literature on quantum foundations for three quarters of a century have given way to new discoveries. As attention gradually shifted from interpretative questions to the engineering challenge of harnessing quantum physics to perform information processing tasks, entanglement and non–locality, once considered an embarrassment, were now seen as a resource. Moreover, the promise of the new engineering endeavor has prompted many to use it as a new framework for conceptual foundations (59). Attempts to reconstruct quantum theory on the basis of information–theoretic principles saturate the literature (40), and the long standing puzzles of non–locality and the measurement problem are mocked as archaic and unworthy (60).

Agreed, there is a certain pedagogical merit in describing the mathematical structure of quantum physics, as well as how and where it differs from the structure of classical physics, using information–theoretic jargon; at this day and age, students are probably more appreciative of such an exposition than, say, of reading the dry language of von Neumann's masterpiece from 1932. And yet the information–theoretic approach, in its attempt to dissolve the old puzzles in the foundations of quantum theory, has generated puzzles of its own that stem from unwarranted double standards. In this final part, we shall focus on two such cases that arise in the context of quantum error correction and which reflect a great deal of confusion that can be traced to an underlying methodological lacuna among quantum information scientists, namely, the view that quantum theory should be considered fundamental *no matter what*.

4.2 NOISE, CLASSICAL OR QUANTUM?

Our discussion of active error correction and the threshold theorems has already revealed a case of double standard in the quantum information literature. Recall that the first models underlying the beautiful mathematical results of these theorems were based on Markovian noise. To repeat, the assumption of Markovian noise means that while the QIP device evolves according to the quantum dynamical laws, its qubits being entangled along the way, no entanglement between the errors during the interaction with the QIP device is allowed; the noise was seen as acting on the qubits *independently* of the structure of the evolution of the QIP device. In such a case, the evolution only *propagates* the external errors, and the latter remain *local*, in contrast to the behavior of the qubits *inside* the QIP device.

The assumption of Markovian noise did allow the mathematicians to calculate feasible thresholds, but apart from this, there was no physical justification for such a double standard. In a truly physical situation, the environment keeps track of the QIP evolution and "learns" it. In such a case, the dynamics not only propagates the errors, but also entangles them, giving rise to non–local noise.

This double standard, namely, treating the system as "quantum" and its environment as "classical", is reminiscent of the famous Copenhagen interpretation. The idea that in order to make sense of quantum theory (be it non–relativistic or relativistic) one must always maintain a classical residue in one's physical description, is pertinent to the writings of Bohr, Heisenberg, and Rosenfeld (121). It is thus not surprising that one finds it in quantum information theory, which, if one believes the words of its own advocates, is nothing but a modern continuation of the Copenhagen tradition (116). What *is* surprising, however, is the confusion that this double standard has generated in the debate on the feasibility of large scale, fault–tolerant, and computationally superior QIP devices.

4.2.1   COMMIT A SIN TWICE AND IT WILL NOT SEEM A CRIME

Quantum information scientists dismiss skepticism about the feasibility of large scale, fault–tolerant, and computationally superior QIP devices by arguing that, given 'proofs of concept' such as the threshold theorems, the obstacles in the quest for building such a machine are merely technological. Consequently, whatever skepticism one might have with respect to this issue must be motivated by one's ideology about quantum theory, namely, one's reluctance to see it as a fundamental theory in physics, that applies universally to all physical scales. (1).

Note, however, that when one analyzes the premises behind this dismissal, one finds that it is the quantum information scientists who require the quantum description to stop short of universality. As stressed in this monograph, the proofs of concept upon which the original optimism about the feasibility of large scale, fault–tolerant, and computationally superior QIP devices was erected presupposed *classical* noise. If one allows the noise to be *quantum*, that is, if one allows the errors to evolve non–locally as the qubits inside the QIP device do, then the jury is still out with respect to the feasibility question. In other words, it is the skeptics and not the believers who demand a universal applicability of the formalism of quantum theory to the dynamical evolution of *both* the QIP device *and* its (arguably much larger) environment.

Quantum information scientists who thus dismiss as ideological skepticism about the feasibility of large scale, fault–tolerant, and computationally superior QIP devices are guilty of a double lapse. Not only do they restrict the universality of quantum theory by treating the QIP device and its environment differently, they also accuse their opponents of committing their own sin.

4.3   THE RETURN OF MAXWELL'S DEMON

A second double standard that the debate on the feasibility of large scale, fault–tolerant, and computationally superior QIP devices has revealed concerns the way in which quantum error correction is depicted in the quantum information literature. Here is a typical text:

Quantum error correction may be thought of as a type of a refrigeration process, capable of keeping a quantum system at a constant entropy, despite the influence of noise processes which tend to change the entropy of the system ... Quantum error correction is essentially a special type of Maxwell's demon — we can imagine a 'demon' performing syndrome measurements on the quantum system, and then correcting errors according to the result of the syndrome measurement. Just as in the analysis of the classical Maxwell's demon, the storage of the syndrome in the demon's memory carries with it thermodynamic cost, in accordance with Landauer's principle. ((119) p. 569).

There are essentially two basic questions here: (a) the physical possibility of Maxwell's demon and (b) the merit of the arguments, based on Landauer's principle, for its exorcism. Both these questions are still open (50), and it is not my intention to go into details here. My modest aim is only to point at another confusion that surrounds the threshold theorems. For this reason, I shall assume neutrality with respect to questions (a) and (b) above, and focus on the claims made in the quantum information community:

I. Quantum error correction (QEC) is essentially a type of Maxwell's demon, namely, a physical situation in which thermodynamics breaks down.

II. Maxwell's demon is subject to Landauer's principle, which, if true, restores the applicability of thermodynamics to the QIP device (plus the QEC gadget plus their environment).

III. The threshold theorems tell us that, below a certain noise threshold, arbitrarily long resilient quantum computation can take place on arbitrarily large input, i.e., QEC can go on forever on an increasingly large QIP device *with essentially no thermodynamic cost* in the total system (QIP plus QEC plus their environment).

Clearly, one cannot consistently hold the above three statements together. To see why, recall that Maxwell's demon was originally conceived to "pick a hole" in the second law of thermodynamics, i.e., to emphasize its statistical character and, therefore, its limited applicability (105). On the other hand, Landauer's principle was an attempt to impose the applicability of thermodynamics even in cases were it supposedly fails to apply (120). Given that we agree, for the sake of the argument, with the quantum information community that quantum theory is fundamental, and that decoherence gives rise only to the "appearance" (for all practical purpose) of a classical world in the macroscopic regime (166), we can now ask the following: are the QEC gadgets subject to the laws of quantum theory or are they subject to the laws of thermodynamics?

Note that this is not an easy question. If the QEC gadgets obey the laws of thermodynamics (through Landauer's principle), then they seem to defy the threshold theorems hence (according to quantum information scientists) to defy the universality of quantum theory and its applicability to the macroscopic regime. If, on the other hand, they obey the laws of quantum mechanics and defy decoherence, then they seem to violate Landauer's principle (which, to emphasize, is nothing but a restatement of the second law of thermodynamics). Consequently, if an unqualified version of (III) is to be defended, then one cannot hold both the universality of quantum theory and the universality of thermodynamics; one of them must give. The quantum information community, however, is ready to give neither and is thus led into an impasse.

Suppose, as the combinatorial approach (and the discussion in chapter 3) suggest, we mitigate statements (II) and (III) and replace them with the following:

IV. The applicability of thermodynamics is not universal, i.e., below a certain threshold (in our case, the size of the QIP device), physical systems can violate it and behave thermodynamically abnormal.

V. Depending on the actual noise level, limited resilient quantum computation can take place on a limited input size, with the QIP device losing its computational superiority as its input increases.

Contrary to statements (II) and (III), these weaker versions do lead one out of the impasse: allowing a limited amount of quantum computation to take place for a limited amount of time, they both are consistent with (I) — the statement that thermodynamics may break down in the microscopic scale.

Since the question whether (V) holds is an *empirical* one, it seems safer to abandon an *a priori* principle such as Landauer's than to face contradiction with facts. This is, essentially, the lesson of this monograph, that suggests a skeptical conjecture about the feasibility of large scale, fault–tolerant, and computationally superior QIP devices, one which acknowledges their physical possibility (thereby accepting (I) and replacing (II) with (IV), thus rejecting the universality of thermodynamics) while at the same time admits their limited scalability (by replacing (III) with (V)). But quantum information scientists cite Landauer's principle (II) as one of the motivations of the entire quantum information industry (23), and it is hard to see them abandoning it after all these years (24).

The methodological rigidity within the quantum information community towards quantum theory leads to an impasse in the context of quantum error correction, as it seems to be irreconcilable with yet another rigidity prevalent in that community towards thermodynamics. But this view of quantum theory as fundamental *come what may* has also led that community to another impasse, which, from the broader perspective of the philosophy of science, is even more disturbing.

4.4   PROGRESS, OR LACK THEREOF

The term *experimental metaphysics* was coined by Abner Shimony to designate the remarkable chain of events that led from Bohm's re–formalization of the EPR argument, through Bell's theorem, to Aspect's experiments. It is well known that by focusing on one of the major problems in the foundations of quantum theory, namely the completeness of the theory, the protagonists of this chain of events exposed the price of holding to any hidden variables theory as a viable alternative.[1] It is also noteworthy, however, that in the discussion on yet another foundational problem, namely

---

[1]What is less appreciated is that quantum theory itself, and indeed *any* theory that reproduces its predictions, was proven by Bell to be non–local.

the problem of measurement, this price was not the only weapon used against the alternatives to quantum theory.

Indeed, being empirically indistinguishable from standard quantum theory, *phenomenologically* local hidden variables theories such as Bohmian mechanics (where the equilibrium distribution of particles' position obeys the Born rule and signaling is not allowed) are often criticized as 'bad science', or, to use Lakatos' famous term, as 'a degenerate research program'. The common claim is that these theories invoke unobserved entities, similar to the Lorentzian ether, and a conspiratorial dynamical mechanism that forever hides these entities, and do so, moreover (and this is the crucial point) in an *ad hoc* way, solely for the sake of maintaining empirical indistinguishability from quantum theory. A telling example for a criticism of this sort can be found in letter written by Steven Weinberg to Sheldon Goldstein in 1996:[2]

> At the regular weekly luncheon meeting today of our Theory Group, I asked my colleagues what they think of Bohm's version of quantum mechanics. The answers were pretty uniform, and much what I would have said myself. First, as we understand it, Bohm's quantum mechanics uses the same formalism as ordinary quantum mechanics, including a wave function that satisfies the Schrödinger equation, but adds an extra element, the particle trajectory. The predictions of the theory are the same as for ordinary quantum mechanics, so there seems little point in the extra complication, except to satisfy some a priori ideas about what a physical theory should be like. . . In any case, the basic reason for not paying attention to the Bohm approach is not some sort of ideological rigidity, but much simpler — it is just that we are all too busy with our own work to spend time on something that doesn't seem likely to help us make progress with our real problems.

Weinberg is not alone in his complaint against indistinguishable alternatives to orthodox quantum theory. Similar opinions can be also found among philosophers and physicists who advocate an information–theoretic approach to the foundations of quantum theory (37) (59). Inspired by the rise of quantum information technologies, in the basis of this approach, lies an ontological view that is succinctly put in a recent letter to *Nature* by one of the most renowned experimentalists in this domain and in a pedagogical article on quantum computing whose author is yet another prominent physicist:

> We have learned in the history of physics that it is important not to make distinctions that have no basis — such as the pre–newtonian distinction between the laws on Earth and those that govern the motion of heavenly bodies. I suggest that in a similar way, the distinction between reality and our knowledge of reality, between reality and information, cannot be made. There is no way to refer to reality without using the information we have about it. ((165) p. 743)

---

[2]The letter, from September 24 1996, is reprinted in (74).

The quantum state of a system is not an objective property of that system. It merely provides an algorithm enabling one to infer from the initial set of measurements and outcomes (state preparation) the probabilities of the results of a final set of measurements after a specified intermediate time evolution. We ourselves have direct access to nothing beyond the outcomes of such measurements. ((116) p. 59)

Apart from this 'epistemic' view of quantum theory, what proponents of the information–theoretic approach share is an attitude towards the quantum measurement problem that regards it as a problem about people — those people who have not seen the light, and still do not understand that quantum theory is not about the world, but about information, or our knowledge of the world (60). Here Bohmian mechanics must be rejected not only because it is empirically indistinguishable from orthodox quantum theory given that certain information–theoretic constraints hold in our world (36), but because, on final account, it is an *ad hoc* solution to a non–problem.

Clearly, that *one* possible solution to the quantum measurement problem is empirically indistinguishable from quantum theory doesn't entail that *all* possible solutions are so indistinguishable. In fact, another class of possible solutions to this problem modifies Schrödinger's equation hence yields *different* predictions than quantum theory in certain situations. According to these collapse theories, entangled states become dynamically unstable as the quantum system at hand becomes macroscopic; hence, we shouldn't expect them to exist in macroscopic systems; quantum theory, on the other hand, predicts that entangled states exist regardless of whether the system at hand is macroscopic or microscopic. Consequently, and contrary to hidden variables theories, collapse theories and quantum theory disagree on what appears to be a purely empirical matter, namely the existence or inexistence of macroscopic entangled states.

Why, however, we do not see too many macroscopic entangled states around us? To answer this embarrassing question quantum theory invokes what has recently been called *the new orthodoxy* (35), namely, environmentally induced decoherence. According to this orthodoxy, macroscopic entangled states exist, but for all practical purposes remain unobserved because we have no control over the states of the environment, so that the reduced state of a decohering system is practically indistinguishable from a classical mixture.

It has been argued (72) that by invoking environmentally induced decoherence, the new orthodoxy renders quantum theory a theory that cannot be caught in telling a lie. Unfortunately, one person's insult is another person's compliment (125). Here another argument will be presented that exposes the double standard that is latent in the new orthodoxy. This double standard is part of a more general drawback in recent attempts to rely solely on decoherence as a (dis)solution to the quantum measurement problem.

The argument is the following. Proponents of this dis(solution) of the measurement problem mock hidden variables theories as *ad hoc* and as exemplifying 'bad science'. Yet by refusing to acknowledge that the measurement problem is a genuine problem that admits an empirical solution, they are willing to tolerate exactly the same *ad hoc*ness with respect to their own theory, hence commit the very sin they find in hidden variables theories. Consequently, they must either renounce their criticism of hidden variables theories (which, as in the case of abandoning Landauer's principle, is an unlikely option) or admit that they also are holding on to a degenerate research program.

4.4.1   TRUE AND FALSE COLLAPSE

In order to set the stage, it is instructive to reconstruct the quantum measurement problem as a problem about *un*observed predictions. This problem arises from a conjunction of a philosophical stance and an indisputable fact. The philosophical stance is taking the formalism of our theories seriously as representing our world. The indisputable fact is that certain predictions made by our theories are never observed. Indeed, if one regards the formalism of quantum theory to be complete and universal, then one must acknowledge that it should apply also to macroscopic massive measurement apparatuses, yet while "electrons enjoy the cloudiness of waves", macroscopic pointers are always localized.

Portrayed in this way, the quantum measurement problem bears similarity to another famous problem in the foundations of classical SM. In thermodynamics, the approach of physical systems towards equilibrium is ubiquitous, and yet the time–reversal–invariant character of the underlying dynamical laws which presumably govern thermal phenomena equally allow fluctuations away from equilibrium to occur. The fact that these are rarely reported is commonly known as the puzzle of the thermodynamic arrow in time.

Initially, there exist three possible solutions to the general problem of unobserved predictions that can be nicely distinguished according to the relevant modality of the predicted (yet *un*observable) phenomena. The first and easiest solution puts the blame on our models. It suggests that we should devise new ones in which such phenomena would be forbidden by law. The second puts the blame on us. We are not clever enough, and we lack the appropriate technological capacities to observe such phenomena, our macroscopic observables being so crude that in appropriate time scales they can be described as evolving autonomously in a single time direction. The third puts the blame on the rarity of the micro–conditions that generate unobserved phenomena Although these are indeed possible, they are also highly improbable, and we are very unlikely to run into them by chance.

How the founding fathers of SM divided these three possible solutions among themselves is well known. What I'd like to suggest here, however, is that a similar division of labor occurs today in the foundations of quantum theory with respect to the measurement problem. Collapse theories follow the first route and modify Schrödinger's equation, thus turning entangled states in macroscopic systems unstable hence nonexistent, while the new orthodoxy invokes environmentally induced decoherence and views macroscopic entangled states as existent yet *practically* unobserved. As mentioned in chapter 3, it has recently been shown (129) how one can follow Boltzmann's insight in his solution of the puzzle of the thermodynamic arrow in time and take the third route, claiming that under a certain assumption of equi–probability imposed not on the space of initial conditions but on the space of conceivable experimental setups, operators that can serve as entanglement witnesses become rare as the number of degrees of freedom of the system at hand increases.

Here I would like to set aside the discussion on the merit and plausibility of the combinatorial approach that characterizes Boltzmann's and Pitowsky's explanations of unobserved predictions in SM and quantum theory, respectively, and rather focus on the other two possible approaches to this problem, which in the context of quantum theory are embodied in true and false collapse (123). In order to appreciate the empirical character of this problem, it is sufficient to demonstrate that these two approaches yield *different* predictions in certain circumstances, and as such, can be distinguished *in principle*.[3]

Suspending discussion on these experimental circumstances to the next section, in what follows, I shall confine myself to a general argument for this distinguishability. For the sake of brevity, I will adopt here the recently promoted quantum–information–theoretical outlook.[4] According to an influential theorem proved by Clifton, Bub, and Halvorson (40), one can characterize quantum theories with the help of three information–theoretic principles, namely, no signaling, no broadcasting, and no (unconditionally secured) bit commitment (*no–bit* henceforth). These three principles serve to 'filter–out' from the more general mathematical structure of C^*–algebra the algebraic structure of operators and states that characterize quantum theory, hence capture the basic aspects of the quantum–theoretic description of physical systems. That is, these three principles serve among others things as constraints on any alternative dynamical theory that aims to reproduce the predictions of quantum theory: if (and only if) one of these principles is violated by the alternative theory, this theory must differ in its predictions from quantum theory.

How do these principles capture the basic aspects of the quantum–theoretic description of physical systems? For a composite system A+B, the no signaling constraint entails that the C^*–algebras $\mathfrak{A}$ and $\mathfrak{B}$, whose self-adjoint elements represent the observables of A and B, commute with each other (this feature is sometimes called *micro-causality*); and the no broadcasting constraint entails that each of the algebras $\mathfrak{A}$ and $\mathfrak{B}$ is noncommutative. Thus, the quantum mechanical phenomenon of interference is the physical manifestation of the noncommutativity of quantum observables or, equivalently, the superposition of quantum states.

The status of the third principle (the *no bit* principle) is somewhat more involved. Some (152) argue against Clifton, Bub, and Halvorson that this principle is actually entailed by the first two,[5] and raise the following dilemma: if the third principle is not redundant, then either it serves to 'rule–in' certain states as *physical* states and not as mere artifact of the formalism (which is conceptually

---

[3]Many experimental scenarios in condensed matter physics were conceived under a research program advocated by A.J. Leggett since 1980, proposes to test the validity of quantum theory in the macroscopic regime. This research program has resulted in beautiful experiments and discoveries of new phenomena, and shall be reviewed in section (4.4.2). In recent philosophy of physics literature, there is one scenario, namely the spin echo experiment, that has been extensively discussed, especially in the context of the foundations of SM (9) (81) (82).

[4]It is important to appreciate that the information–theoretic outlook (37), whatever merit it may carry, simply re–describes this distinguishability, that follows from two simple facts: (1) that the predictions of quantum theory are given by Born's probability rule, and (2) that true collapse theories *modify* this rule.

[5]In two separate systems whose algebras are non-commutative, there are pure states which are not product states, i.e., one gets entanglement just from no signaling and no broadcasting. Note that this proof relies on the premise that the notion of C^* algebraic independence and the notion of kinematic independence (the first says that any two states in two separated systems can be combined to form a state with the given states as marginals; the second says that two algebras of the two systems mutually commute) are jointly equivalent to the notion of tensor product (18).

dubious given the contrary *restrictive* role the other two principles) or it serves to capture a dynamical feature of quantum theory, rather than a kinematical one, as claimed by Clifton, Bub, and Halvorson.

But the role of the *no bit* principle is to restrict a type of theories that are quantum mechanical in their kinematics (hence impose no kinematical — or structural — superselection rules), yet involve different *dynamics* than quantum theory which in turn are responsible for the decay of entangled states in certain circumstances.[6]

Be that as it may, it is interesting to find a curious remark about the role of the no–bit principle as an effective principle that can decide between quantum theory and rival alternative theories ((152) p. 207):

> Note, though, that a scenario could be imagined in which the no bit–commitment condition would play more of an active role. If, for some reason, we were unsure about whether a Schrödinger–type theory or a quantum theory were the correct physical theory, then being informed by an oracle whether or not unconditionally secure bit–commitment was possible would be decisive: we would be saved the effort of having to go out into the world and perform Aspect experiments. But as this is not our position, the no bit–commitment axiom does not play an active role in picking out quantum theory.

I find this remark mistaken in two distinct senses. First, it is clear that its author, referring as he does to the Aspect experiment and to the alleged redundancy of the *no bit* principle, has in mind only the distinction between quantum and *classical* theories. But, of course, Schrödinger–type theories are *non*–classical at least up to the moment in which entanglement decays. Indeed, the correct role of the *no bit* principle is not to distinguish between *classical* and quantum theories but to 'filter–out' exactly those *quantum* theories whose dynamical law differs from quantum theory and leads to decay of entanglement. This remark, in fact, *ignores a priori* the possibility of collapse theories as a viable alternative to quantum theory. This leads its author to his second mistake: that one must *consult an oracle* in order to know whether this principle is violated or not in our world. But as a matter of fact, one has at one's disposal an arsenal of possible, legitimate theories, a whole class of spontaneous localization theories, or collapse theories such as the GRW and CSL theories (124), which are exactly of the type of theories whose dynamics lead to the decay of entangled states in precisely described scenarios.

The GRW theory (formulated for non–relativistic quantum mechanics) explains the *un*observability of some macroscopic superpositions of *position* states by modifying the Schrödinger linear dynamics in such a way that given the new dynamics such superpositions are overwhelmingly likely to collapse at every moment of time, and in this sense, they are highly unstable. The Schrödinger equation is changed by adding to it a non-linear and stochastic term that induces the so-called *jump* or collapse of the wavefunction. The jump is supposed to occur on occasion in position

---

[6]A theory of this type that was conceived by Schrödinger in 1936 (140), where entanglement decays with the increase in spatial distance between the separated systems. Schrödinger's idea has been reconsidered by Furry and by Aharonov and Bohm, and was finally shown to be untenable in (70) (since it doesn't even admit a statistical operator description and would require a complete reformulation of quantum theory). In the current context, collapse theories (71) (122) serve as yet another example of Schrödinger–type theories.

space and its postulated frequency is proportional roughly, to the mass density of the system (or in Bell's model (22) on the number of particles described by the wavefunction). For our purposes, it is enough to sketch Bell's version of the elementary and non–relativistic theory of GRW. This goes roughly as follows.

Consider the quantum mechanical wavefunction of a composite system consisting of N particles:

$$\psi(t, \mathbf{r}_1, \mathbf{r}_2, ..., \mathbf{r}_N). \tag{4.1}$$

The time evolution of the wavefunction usually (at almost all times) satisfies the deterministic Schrödinger equation. But sometimes *at random* the wavefunction collapses or *jumps*) onto a wavefunction ψ_ℓ localized in position of the (normalized) form

$$\psi_\ell = \frac{j(\mathbf{x} - \mathbf{r}_n)\ \psi(t, \mathbf{r}_1, \mathbf{r}_2, ..., \mathbf{r}_N)}{R_n(\mathbf{x})}, \tag{4.2}$$

where $\mathbf{r}_n$ in the jump factor $j(\mathbf{x} - \mathbf{r}_n)$ (which is normalized) is randomly chosen from the arguments $\mathbf{r}_1, ..., \mathbf{r}_n$ of the wavefunction immediately before the jump, and $R_n(\mathbf{x})$ is a suitable renormalization term. For j, GRW suggest the Gaussian:

$$j(\mathbf{x}) = K\ \exp(-\mathbf{x}^2/2\Delta^2), \tag{4.3}$$

where the width Δ of the Gaussian is supposed to be a new constant of nature: $\Delta \approx 10^{-5}$cm.

Probabilities enter the theory twice. First, the probability that the *collapsed* wavefunction ψ_ℓ after a jump is centered around the point $\mathbf{x}$ is given by

$$\mathrm{d}^3\mathbf{x}\,|R_n(\mathbf{x})|^2. \tag{4.4}$$

This probability distribution, as can be seen, is proportional to the standard quantum mechanical probability given by the Born rule for a position measurement on a system with the wavefunction $\psi(t, \mathbf{r})$ just prior to the jump. Second, the probability in a unit time interval for a GRW jump is

$$\frac{N}{\tau}, \tag{4.5}$$

where N is the number of arguments in the wavefunction (i.e., in Bell's model it may be interpreted as the number of particles), and τ is, again, a new constant of nature ($\tau \approx 10^{15}$ sec $\approx 10^8$ year). Note that the expression (4.5) does not depend on the quantum wave function, but only on N. This is essentially the whole theory.

Theories such as GRW or CSL are compatible with everything we know so far about the world (hence do not violate the no–bit principle in *micro*scopic systems) but still yield *different* predictions than standard quantum theory hence presumably violate this principle in *macro*scopic systems.[7]

---

[7]For microscopic systems, GRW collapses have extremely low probability to occur, so that the quantum mechanical Schrödinger equation turns out to be literally true at almost all times in just the way that no collapse quantum mechanics predicts (and

Note, however, that in such *macro*scopic scenarios standard quantum theory also predicts an *effective* violation of the no–bit principle in the same sense due to environmental decoherence (i.e., such states effectively collapse also in the standard theory as all models of environmentally induced decoherence show). So practically, quantum theory and the GRW theory agree on the *no bit* principle for all cases in which there are good empirical reasons to believe it is true. And the two theories seem to disagree about the *no bit* principle only with respect to those macroscopic superpositions about which, given the problem of unobserved predictions, we do not know whether or not they in fact exist in our world. But, of course, this is no surprise since this disagreement is located precisely where the two theories differ in their empirical predictions. So we are back to square one![8]

It thus seems that the kind of experimental metaphysics the quantum–information–theoretic approach is eager to avoid is nevertheless the natural consequence of this state of affairs, which is succinctly captured by Clifton's, Bub's, and Halvorson's characterization theorem. In this context, however, we are not trying to distinguish anymore quantum theory from hidden variables theories as Bell and Aspect did, but rather true from false collapse. That *this* (and not the dismissal of the measurement problem) should be the philosophical significance of the Clifton–Bub–Halverson theorem is argued in length in (77). Philosophical arguments aside, what makes the quantum–information–theoretic outlook even more puzzling is the fact that the very type of experimental metaphysics it seems to dismiss has been *actually taking place* for the last 30 years in the lively domain of condensed matter physics.

4.4.2   THE SEARCH FOR QIMDS

Over the last 30 years or so, condensed matter physics has been witnessing a realization of a research program which is the exact type of experimental metaphysics that quantum–information–theorists have been emphatically ignoring. As one on the initiators of this research program, the Nobel laureate A.J. Leggett, puts it in a review paper that summarizes the motivation of this program and its prospects:

> Perhaps fortunately for the topic reviewed in this paper, experimental physicists tend not to be avid or credulous readers of the quantum measurement literature, and the skepticism expressed in much of that literature as to the possibility of ever observing QIMDS [quantum interference of macroscopically distinct states - AH] has not prevented groups in several different sub–fields of physics from tying experiments in this direction. ((108) p. R432)

experiment confirms). However, for massive macroscopic systems (e.g., for systems with 10^{23} particles) the GRW collapses are highly probable at all times. In measurement situations, the GRW theory implies that superpositions of macroscopically distinguished pointer states collapse with extremely high probability onto the localized states on time scales that are much faster than measurement times. In particular, the probability that the wavefunction of the composite of system plus apparatus will stay in a superposition for more than a fraction of a second (e.g., by the time the measurement is complete) vanishes exponentially.

[8]Note that the point made here is quite general: it would be applicable to *any* characterization of the mathematical structure of quantum theory in terms of information–theoretic principles and to any dynamical alternative of it, provided the latter is empirically well confirmed and *not* equivalent to standard quantum theory.

As a result of these experimental efforts,[9] and considerable skepticism about the prospects of ever observing quantum macroscopic superpositions involving more than a few "elementary" particles notwithstanding, progress in the quest for quantum interference of macroscopically distinct states has been spectacular. Ranging from traditional Young's slits experiments conducted with C_{60} and C_{70} molecules (15) to SQUID experiments in which the two superposed states involved $\sim 10^{10}$ electrons behaving differently (57), these results are beginning to impose non–trivial constraints on possible collapse theories.

Initiating this research program, Leggett (106) has proposed a precise quantitative measure for 'macroscopically distinct' quantum states in terms of various reduced density matrices, which will allow classification thereof along the 'axis' that takes one from the well–verified single–particle interference experiments of the Young's slits type to the paradoxical situations of which Schrödinger's cat is paradigmatic.[10] One could then attach different degrees of this measure to different experimental set–ups, with the aim of verifying whether the macroscopically distinct quantum states predicted by quantum theory are observed in these set–ups.

The fundamental idea of Leggett's research program predates collapse theories such as GRW and CSL. In the latter, the quantitative measure for macroscopic distinctiveness is related to the difference in displacement of the system's centre of mass in the two branches of the relevant wave function. Unfortunately, although they do impose some stringent constraints on alternative collapse theories, most of the experiments conducted to date, conceived under Leggett's program, are insufficient to refute GRW and CSL.[11] One might object that there is no *a priori* physical motivation for choosing the displacement of the centre of mass as a measure of macroscopic distinctiveness. and that one should not regard GRW and CSL as exhausting, or even as making major inroads into, the possibilities for physical collapse theories. I agree, yet nevertheless, there are independent *philosophical* motivations for dynamical reduction theories such as GRW and CSL.

Metaphysically, there is nothing in these theories but the wave function, hence their ontology and the meaning they attributes to the quantum probabilities, given as they are by the (modified) Born rule, are both transparent and precise — a feature already highlighted by J.S. Bell (22). Moreover, from a methodological point of view, while these theories involve no external triggers (since the collapse they introduce is spontaneous), they are still phenomenological in the sense that they leave open the possibility of incorporating new physics from the intersection of QM and the general theory of relativity, into their models. From this perspective, the choice of the centre of mass variable in these models, as opposed to equally possible macroscopic 'triggers', while lacking any *a priori* motivation, is entirely understandable and natural. Finally, the dynamical reduction program has, by far, produced the most complete and 'worked–out' collapse models than any other 'macro–realistic'

---

[9]Ironically, most of these results were also motivated by the search for a large scale quantum computer.

[10]Leggett's measure, coined *disconnectivity*, appears to be essentially equivalent to the notion of 'degree of entanglement,' which has received a lot of attention in the context of generalized Bell–type inequalities ((107) p. 8).

[11]Note the methodological difference between Leggett's program, which aims for validating quantum theory in the macroscopic realm, and the dynamical collapse program, which aims to refute it. Indeed, for the dynamical collapse program to succeed, one should look for experiments that expose deviations from standard quantum theory predictions and should not be content with just keeping the dynamical collapse program unfalsified by current experiments. For recent progress see (3) (4) (17).

alternatives to quantum theory. Moreover, in recent years considerable progress has been made in generalizing it to the relativistic domain (153).

Be that as it may, since the argument to which I now turn is, on final account, quite generic, I would like to emphasize again that nothing in the ensuing discussion hinges upon one's choice in a particular collapse model.

4.4.3 THE *ANCILLA* ARGUMENT

We are now ready to appreciate the double standard behind the information–theoretic attitude towards Bohmian mechanics and the measurement problem. Recall that this viable alternative to quantum theory is mocked as 'bad science' and 'degenerate research program'. Yet when it comes to lagging behind the phenomena and advancing *ad hoc* modifications of one's theory, come what may, the proponents of the new orthodoxy are in the forefront, as the following discussion will now show.

Suppose that crucial experiments similar to the ones that are being carried out under Leggett's research program that are capable of distinguishing between, say, the GRW theory and environmentally induced decoherence were to come out in accordance with the GRW predictions to a very good approximation. Nevertheless, it is often argued that standard quantum theory would remain *intact* for the following reason.

Let us suppose that an *open* system S is subjected to *perfect* decoherence, namely to interactions with some degrees of freedom in the environment E, such that the states of the environment become strictly orthogonal. Suppose further that we have no access whatsoever (as a matter of either physical fact or law) to these degrees of freedom. In this case, the GRW dynamics for the density operator of S would be indistinguishable from the dynamics of the *reduced* density operator of S obtained by evolving the composite quantum state of $S + E$ *unitarily* and tracing over the inaccessible degrees of freedom of E. It turns out that this feature is mathematically quite general because the GRW dynamics for the density operator is a completely positive linear map ((119) pp. 353–373, (148) fn. 14). From a physical point of view, this means that the GRW theory is empirically equivalent to a quantum mechanical theory with a unitary (and linear) dynamics of the quantum state defined on a *larger* Hilbert space. In other words, one could always introduce a new quantum mechanical ancilla field whose degrees of freedom are inaccessible, and one could cook up a unitary dynamics on the larger Hilbert space that would simulate the GRW dynamics on the reduced density operator. Therefore, experimental results that might seem to confirm GRW–like dynamics could always be *re–interpreted* as standard quantum theory on larger Hilbert spaces. In particular, such a theory could always be made to satisfy the three information–theoretic constraints suggested by Clifton, Bub, and Halvorson (40).[12]

Admittedly, some quantum–information theorists will argue that at our current level of quantitative understanding of condensed matter systems, any prima facie failure to see superpositions

---

[12]Note that although the ancilla theory could always be made to satisfy the three information–theoretic principles of Clifton, Bub, and Halvorson (in particular the *no bit* principle) on the larger Hilbert space, unconditionally secure bit commitment would be possible, in so far as our experimental capacities are concerned, via protocols that require access the ancilla field (which *ex hypothesi* is inaccessible). So, the *ancilla* argument renders the *no bit* principle as a constraint on the feasible flow of information quite idle.

where we expected to, rather than being evidence for GRW or CSL, is most likely to be due to currently unidentified sources of decoherence. Advocates of this point of view would presumably be willing to concede that with improved experimental data and/or theoretical arguments (39), it may, at some stage, be necessary to abandon this loophole and concede that something like GRW or CSL is correct; they just do not feel that this stage has been reached to date.

But other advocates of the quantum–information approach to the foundations of quantum theory, dismissive as they are of the measurement problem and supporting the above *ancilla* argument as they do, seem to hold the astonishing position that quite irrespective of detailed theoretical and experimental considerations, *any* failure to see superposition where expected should *automatically* be attributed to unknown (and perhaps never–to–be–known) sources of decoherence. Yet clearly in the present state of quantum theory, we are far from being able to claim — as the above *ancilla* argument does — that the theory must be protected *come what may*. In fact, it is not even clear what such an attitude might mean in even the weakest form of an empiricist approach to theoretical physics. The ancilla field in the above argument has, by construction, no observable effects (44), and this amounts to introducing hidden variables (or more appropriately, a new 'quantum ether') into standard quantum theory, whose sole theoretical role is to save some disputable principles against (putative) empirical refutation.

But now, If one is willing to accept *ad hoc* such an argument in the context of quantum theory, then why should one reject similar 'ether'–like approaches in the context of relativity theory or hidden variables theories such as Bohmian mechanics? Such approaches are sweepingly rejected (as Weinberg's letter demonstrates) mainly because their complex underlying structure doesn't translate into new empirical predictions. It thus seems that by propounding the *ancilla* argument, the proponents of quantum theory, who criticize Bohmian mechanics as 'bad science', render themselves utterly inconsistent.

Arguably, one *might* have good reasons to protect unitary quantum theory against what might seem as straightforward empirical refutation. From a theoretical point of view, it might turn out that both collapse and hidden variables theories could not be made compatible with some fundamental physical principles that we cannot give up without giving up some significant chunk of contemporary theoretical physics (conservation of energy or Lorentz–covariance might be such examples).[13] In that case, a protective argument such as the *ancilla* argument might be understandable. But the point is that the present state of quantum theory doesn't warrant such an argument. This is mainly because quantum theory itself has quite deep foundational problems not only at its most basic level (i.e., the measurement problem) but also, for example, in its generalizations to both special and general relativity. In such circumstances the right epistemological stance is not to ignore alternative collapse theories such as the GRW theory, nor to suspend judgment with respect to them but, instead, to let them stand or fall by empirical confirmation.[14]

---

[13]To be fair, Weinberg in his letter also suggests an additional reason for rejecting Bohmian mechanics, namely its resistance to any extension to quantum field theories.

[14]Leggett (107) (108) suggests that the above *a priori* argument against the feasibility of an empirical solution to the measurement problem can be (and is) blocked by the possibility, given certain plausible assumptions on the microscopic Hamiltonian of the

4.4.4   PROGRESS?

In his "The impossible pilot wave", John S. Bell writes on the dismissive attitude that the 'pilot wave' theory of de-Broglie and Bohm was receiving from the community:

> But why then had Born not told me of this 'pilot wave'? If only to point out what was wrong with it? Why did von Neumann not consider it? More extraordinary, why did people go on producing 'impossibility' proofs, after 1952, and as recently as 1978? When even Pauli, Rosenfeld, and Heisenberg, could produce no more devastating criticism of Bohm's version than to brand it as 'metaphysical' and 'ideological'? Why is the 'pilot wave' picture ignored in the text books? Should it not be taught, not as the only way, but as an antidote to the prevailing complacency? ((22) p. 160)

As is well known, Bell's insistence on pursuing hidden variables theories as a possible solution to the completeness problem of quantum theory led to the first incident of experimental metaphysics. But with respect to the other foundational problem of measurement (for which hidden variables theories are *not* the only solution) he wrote in 1975:

> The continuing dispute about quantum measurement theory is not between people who disagree on the results of simple mathematical manipulations. Nor is it between people with different ideas about the actual practicality of measuring arbitrarily complicated observables. It is between people who view with different degrees of concern or complacency the following fact: so long as the wave packet reduction is an essential component, and so long as we do not know exactly when and how it takes over from the Schrödinger equation, we do not have an exact and unambiguous formulation of our most fundamental physical theory. ((22) p. 51)

More than 30 years have passed, but (quantum information notwithstanding!) we are still far from solving the measurement problem. Agreed, whether or not there exists a problem may depend on one's metaphysical predilection, and condensed matter physicists continue to conduct beautiful experiments regardless of quantum–information theorists' opinions. Yet these opinions, dismissive as they are of collapse theories as they are of the 'pilot wave' view, have rendered quantum–information theorists utterly inconsistent. As the discussion on the *ancilla* argument clearly shows, by deeming the measurement problem 'non–empirical', it is now the quantum–information–theoretic proponents of environmentally induced decoherence who are engaged in a degenerate research program, unable to admit that another round of experimental metaphysics is actually taking place in the physics community today.

system at hand, of making reliable quantum mechanical predictions purely on the basis of knowledge of the classical motion of the system, and remarks that this feasibility (based on (39)) was widely recognized in the relevant community of condensed matter physicists. I strongly agree, but the point I wish to make here is that since the 'ancilla' argument has been propounded over and over again in the quantum–information community, it is quite obvious that at least some portion of the physics community is unwilling to acknowledge the physical possibility of performing an experiment that can detect differences between genuine collapse and environmentally induced decoherence. My only point is that in doing so, this portion of the community displays an unwarranted double standard.

CHAPTER 5

Coda

Since its inception, quantum computing has presented a dilemma: is it reasonable to study a type of computer that has never been built, and might never be built in one's lifetime? Arguments to the negative abound (66) (110) (162) and are based on the claim that there is a fundamental physical (as opposed to practical) reason why large scale (computationally superior) quantum computers can never be built. Yet all these objections fail to present a rigorous challenge to the optimists that would elevate the debate from mere ideology to exact science. Recent years have also taught us that once such a pessimistic challenge is actually presented rigorously, an optimistic response is in the offering, along with innovative and new theoretical and experimental techniques. Thus, quantum error correction was invented as a response to the initial worries about decoherence, and fault-tolerant circuits were conceived as a response to worries about hardware imprecision. The aim of this monograph was to clarify the underlying assumptions behind yet another chain in the link of these challenges that have fueled the quantum computing enterprise.

The question of whether large scale, fault–tolerant, and computationally superior quantum computers could be built is separate from the question of what can be done with them. Indeed, despite knowing nothing about the former, computer scientists have made great progress in the latter. In this final part, I shall argue that the realization question has also interesting *philosophical* consequences for computer science, consequences that go beyond the mere expansion of the abstract landscape of complexity theory.

The limits of computation, both in terms of what problems can or cannot be computed, and in terms of which computable problems can be solved efficiently, were originally conceived within the realm of mathematical logic. Thus, the common formulation of the Church Turing thesis runs as follows: *Every computable function from a subset of the natural numbers to the natural numbers is computable by a Turing machine.* As is well known, Turing and Church were interested in the decision problem for formal systems, and in particular, for arithmetic. They were therefore concerned with a notion of computation that is intimately related to the formal concept of proof. Since formal proofs can be validated, at least in principle, by checking whether each step follows a mechanical rule of inference, computation in the mathematical logic sense should also have that character, i.e., it should be an idealization of symbolic calculation with pencil and paper. Indeed, this is how Turing (154) viewed his machine.

With the advance of technology, computer science left its cradle in mathematical logic and became an independent discipline. This development has also led to a change in the meaning of computation (141), with notions such as "computability" and "complexity" — originally conceived within mathematical logic — migrating into physics. Today we no longer stress proofs and calculations as idealized human activity, but instead cash out these notions in terms of the abilities of a "finite discrete automaton". With this change, some of the transparency of the original Church Turing thesis is gone (131). Attempts were made to clarify this notion by relying on common intuitions (63) (146), but, on final account, the issue is still open. The problem is that what we mean by "finite discrete automaton" is heavily dependent on our spacetime theories, and on what we mean by "physical state". However, even in its contingent form, as an empirical hypothesis, the *physical* Church Turing thesis, namely, that every function that is computable by a finite discrete automaton is computable by a Turing machine, has not yet been seriously challenged (49) (84) (78) (126).

Another way to formulate the physical Church Turing thesis is the following (130): *The distinction between computable and non–computable functions is independent of the machine model*. Trying to gather support for this thesis, it turned out that all the machine models that were computationally equivalent to a Turing machine were of polynomial time complexity (51), and the following independent thesis became more established: *The class P is independent of the machine model* (where P denotes the class of polynomial time Turing computable functions). In other words, no automaton can reduce the complexity of an exponential or a sub–exponential Turing computable function, and compute it in polynomial time.

This *polynomial* Church Turing thesis has so far no counterexample, but there is a serious candidate, namely, Shor's quantum algorithm for FACTORING (143) (assuming, of course, that this problem is not in P and thus admits no polynomial time algorithm).

This putative counterexample bears directly on some long standing questions in the philosophy of mind and cognitive science, e.g., the debate on the autonomy of computational theories of the mind (55). Traditionally, advocates of the computational model of the mind have tried to impose constraints on computer programs before they can qualify as theories of cognitive science (136). Their search for what makes theories *computational* theories of the mind involved isolating some autonomous features of these programs. In other words, they were looking for computational properties, or kinds, that would be machine–independent, at least in the sense that they would not be associated with the physical constitution of the computer, nor with the specific machine model that was being used. These features were also thought to be instrumental in debates within cognitive science, e.g., the debate between functionalism and connectionism (56).

Yet if physical processes, such as, e.g., quantum processes, may re–describe the abstract space of computational complexity, then computational concepts such as complexity classes and even computational kinds such as an efficient algorithm will become machine–dependent, and recourse to hardware will become inevitable in any analysis thereof. Advances in quantum computing may thus militate against the functionalist view about the non-physical character of the types and properties used in computer science. In fact, these types and categories may become physical as a result of this natural development in physics (127).

In sum, the perspective offered in this monograph is likely to be fruitful not only within physics, but also in disciplines such as computer and cognitive science. Let's hope that its impact on these disciplines will be felt long before large scale, fault–tolerant, and computationally superior quantum computers are built.

APPENDIX A

The Dynamics of an Open Quantum System

What follows is based on standard introductions to the theory of Quantum Open Systems (32). In general, an open quantum system S with a Hilbert space $\mathcal{H}_S$ is coupled to another quantum system B, the environment, with Hilbert space $\mathcal{H}_B$ (if B has infinite degrees of freedom, it is sometimes called a 'reservoir', and if the reservoir is in thermal equilibrium, it is denoted a 'heat bath'). Thus, S is a subsystem of the total system $S + B$ living in the tensor product space $\mathcal{H}_S \otimes \mathcal{H}_B$. If the interaction between S and B (with their respective Hamiltonians H_S and H_B) is governed by the Hamiltonian H_I, then the Hamiltonian of the total system can be written as:

$$H = H_S \otimes I_B + I_S \otimes H_B + \alpha H_I, \tag{A.1}$$

where I_S and I_B denote the identities in the Hilbert spaces of the system and the environment, respectively, and α is a coupling constant.

The open system S is singled out by the fact that all observables A of interest refer to this system, and are thus of the form $A \otimes I_B$, where A acts in $\mathcal{H}_S$. If we describe the state of the total system by some density matrix ρ, then the expectation value of the observable A is determined by

$$\langle A \rangle = tr_B\{A\rho_S\}, \tag{A.2}$$

where

$$\rho_S = tr_B\rho, \tag{A.3}$$

is the reduced density matrix (here tr_S and tr_B denote, respectively, the partial traces over the degrees of freedom of the open system S and of the environment B. As the dynamics of the total system is Hamiltonian, the total density matrix evolves unitarily, hence the time–development of the reduced density matrix may be represented in the form

$$\rho_S(t) = tr_B\{U(t,0)\rho(0)U^\dagger(t,0)\}, \tag{A.4}$$

where the initial state of the total system at time $t = 0$ is given by $\rho(0)$ and $U(t,0)$ is the time–evolution operator of the total system over the time interval from $t = 0$ to $t > 0$. The corresponding differential form of the evolution is obtained from a partial trace over the environment of the von Neumann equation,

$$\frac{d}{dt}\rho_S(t) = -itr_B[H(t), \rho(t)]. \tag{A.5}$$

Given that the initial state of the total system is of the form $\rho(0) = \rho_S(0) \otimes \rho_B$, the dynamics expressed through (A.4) can be viewed as a map of the state space of the reduced system which maps the initial state $\rho_S(0)$ to the state $\rho_S(t)$ at time $t \geq 0$,

$$\rho_S(0) \rightarrow \rho_S(t) = V(t)\rho_S(0). \tag{A.6}$$

For a fixed t, this map is known as a 'dynamical map'. Considered as a function of t, it provides a one–parameter family of dynamical maps. If the characteristic time–scale over which the reservoir correlations functions decay are much smaller than the characteristic time–scale of the system's relaxation, it is justified to neglect memory effects in the reduced system dynamics and to expect a Markovian type behavior, which may be formalized by the semigroup property

$$V(t_1)V(t_2) = V(t_1) + V(t_2), \quad t_1, t_2 \geq 0. \tag{A.7}$$

The one–parameter family of the dynamical maps then becomes a quantum dynamical semigroup. Introducing the corresponding generator $\mathcal{L}$, one immediately obtains an equation of motion for the reduced density matrix of the open system S of the form

$$\frac{d}{dt}\rho_S(t) = \mathcal{L}\rho_S(t). \tag{A.8}$$

Such an equation is called a Markovian quantum master equation (MME). The most general form of the generator $\mathcal{L}$ is provided by a number of theorems, e.g., Linblad's (112) according to which:

$$\mathcal{L}\rho_S = -i[H, \rho_S] + \sum_i \gamma_i \left(A_i \rho_S A_i^\dagger - \frac{1}{2} A_i^\dagger A_i \rho_S - \frac{1}{2} \rho_S A_i^\dagger A_i \right). \tag{A.9}$$

Here H is the generator of the coherent part of the evolution (which need not be identical to the system's Hamiltonian) and the A_i are system's operators with corresponding relaxations times γ_i. In a number of physical situations, a quantum master equation whose generator is exactly of the form (A.8), known as the Linblad form, can be derived from the underlying microscopic theory under certain approximations. Two of the most important cases are the singular coupling limit and the weak coupling limit.

The singular coupling limit (SCL) is suitable when one wishes to describe an open quantum system strongly driven by some external macroscopic device with asymptotic states far from equilibrium. To accelerate the decay of the reservoir's correlation, one then re–scales H_B and H_I. In this limit, there is no restriction on the time–dependence of H_S, but the price paid is infinite temperature for the bath. The weak coupling limit (WCL) is much more physical and is obtained for all temperatures, but it requires slower system's dynamics. Both derivations must, of course, be consistent with the constraints of thermodynamics, which in this context are formalized as the Kubo–Martin–Schwinger (KMS) condition (90), namely that the reservoir is considered as a heat bath, i.e., in thermal equilibrium, so that its entropy and temperature could be defined, and that the total system (system + bath) relax to the equilibrium state imposed by these constraints.

APPENDIX B

A Noiseless Qubit

A quantum version of the classical two bit example (where the invariant, conserved, quantity under the total evolution of the system and the errors was the parity) consists of two physical qubits, where the errors randomly apply the identity or one of the Pauli operators to the first qubit. The Pauli operators are defined by

$$\mathbb{I} = \begin{pmatrix} 1 & 0 \\ 0 & 1 \end{pmatrix}, \ \sigma_x = \begin{pmatrix} 0 & 1 \\ 1 & 0 \end{pmatrix}, \ \sigma_y = \begin{pmatrix} o & -i \\ i & 0 \end{pmatrix}, \ \sigma_z = \begin{pmatrix} 1 & 0 \\ 0 & -1 \end{pmatrix}. \tag{B.1}$$

Explicitly, the errors have the effect

$$|\psi\rangle_{12} \begin{cases} \mathbb{I}|\psi\rangle_{12} & \text{Probability .25} \\ \sigma_x^{(1)}|\psi\rangle_{12} & \text{Probability .25} \\ \sigma_y^{(1)}|\psi\rangle_{12} & \text{Probability .25} \\ \sigma_z^{(1)}|\psi\rangle_{12} & \text{Probability .25} \end{cases} \tag{B.2}$$

where the parenthesized superscript (1) specifies the qubit that an operator acts on. This error model is called "completely depolarizing" errors on qubit 1. Obviously, a one–qubit state can be stored in the second physical qubit without being affected by the errors. An encoding operation that implements this observation is

$$|\psi\rangle \rightarrow |0\rangle_1 |\psi\rangle_2, \tag{B.3}$$

which realizes an ideal qubit as a two–dimensional subspace of the physical qubits. This subspace is the "quantum code" for this encoding. To decode one can discard physical qubit 1 and return qubit 2, which is considered to be a natural subsystem of the physical system. In this case, the identification of syndrome and information–carrying subsystems is an obvious one. This example is taken from (99), where the interested reader may find a lucid and accessible introduction to the idea of noiseless subsystem.

Bibliography

[1] S. Aaronson. Multilinear formulas and skepticism of quantum computing. *STOC '04, Proceedings of the 36th annual ACM symposium on Theory of computing*, 118–127, 2004. DOI: 10.1145/1007352.1007378 1, 2, 20, 36

[2] S. Adler. Why decoherence has not solved the measurement problem. http://xxx.lanl.gov/abs/quant-ph/0112095, 2002. DOI: 10.1016/S1355-2198(02)00086-2

[3] S. Adler. Stochastic collapse and decoherence of a non–dissipative forced harmonic oscillator. *Journal of Physics* A, 38, 2729–2760, 2005. DOI: 10.1088/0305-4470/38/12/014 46

[4] S. Adler et al. Towards quantum superpositions of a mirror, An exact open systems analysis—calculational details. *Journal of Physics* A, 38, 2715, 2005. DOI: 10.1088/0305-4470/38/12/013 46

[5] D. Aharonov. Quantum computing. *Annual Review of Computational Physics* VI. World Scientific, 1998. ix

[6] D. Aharonov and M. Ben-Or. Fault–tolerant quantum computation with constant error. http://arxiv.org/abs/quant-ph/9611025, 1996. DOI: 10.1145/258533.258579 1

[7] D. Aharonov et al. Limitations of noisy reversible computation. http://arxiv.org/abs/quant-ph/9611028, 1996. 11

[8] D. Aharonov et al. Fault–tolerant quantum computation with long–range correlated noise. *Physical Review Letters*, 96, 050504, 2006. DOI: 10.1103/PhysRevLett.96.050504 13

[9] D. Albert. *Time and Chance*. Harvard University Press, 2000. 42

[10] R. Alicki. Comments on 'Fault–Tolerant Quantum Computation for Local Non–Markovian Noise'. http://arxiv.org/abs/quant-ph/0402139, 2004. 19

[11] R. Alicki. Comment on 'Resilient Quantum Computation in Correlated Environments, A Quantum Phase Transition Perspective' and 'Fault–Tolerant Quantum Computation with Long–Range Correlated Noise'. http://arxiv.org/abs/quant-ph/0702050, 2007. 20

[12] R. Alicki and M. Fannes. *Quantum Dynamical Systems*. Oxford University Press, 2001. DOI: 10.1093/acprof:oso/9780198504009.001.0001 14

[13] R. Alicki, M. Horodecky, P. Horodecky, and R. Horodecky. Dynamical description of quantum computing: generic nonlocality of quantum noise. *Physical Review* A, 65, 062101, 2002. DOI: 10.1103/PhysRevA.65.062101 19

[14] R. Alicki, D. Lidar, and P. Zanardi. Internal consistency of fault–tolerant quantum error correction in light of rigorous derivations of the quantum Markovian limit. *Physical Review* A, 73, 052311, 2006. DOI: 10.1103/PhysRevA.73.052311 10

[15] M. Arndt. Wave–particle duality of C_{60} molecules. *Nature*, 401, 680–682, 1999. DOI: 10.1038/44348 46

[16] A. Baasi and G.C. Ghirardi. Dynamical reduction models, *Physics Reports*, 379, 257–426, 2003. DOI: 10.1016/S0370-1573(03)00103-0

[17] A. Bassi, S. Adler, and E. Ippoliti. Towards quantum superpositions of a mirror: an exact open system's analysis. *Physical Review Letters*, 94, 030401, 2005. DOI: 10.1103/PhysRevLett.94.030401 46

[18] G. Bacciagaluppi. Separation theorems and Bell Inequalities in algebraic quantum mechanics. In P. Busch *et al.* (Eds.), *Proceedings of the Symposium on the Foundations of Modern Physics 1993.* World Scientific, 1994. 42

[19] A. Barenco, A. Berthiaume, D. Deutsch, A. Ekert, R. Jozsa, and C. Macchiavello. Stabilization of quantum computations by symmetrization. *SIAM Journal of Computing* 26, 1541–1557, 1997. DOI: 10.1137/S0097539796302452 18, 23, 24

[20] J.S. Bell. On the wave packet reduction in the Coleman Hepp model. *Helvetica Physica Acta*, 48, 93–98, 1975. Reprinted in Bell 1987.

[21] J.S. Bell. On the impossible pilot wave. *Foundations of Physics*, 12, 989–999, 1982. Reprinted in Bell 1987. DOI: 10.1007/BF01889272

[22] J.S. Bell. *Speakable And Unspeakable in Quantum Mechanics*. Cambridge University Press, 1987. 44, 46, 49

[23] C. Bennett. Logical reversibility of computation. *IBM J. Res. Dev.* 17, 525–532, 1973. DOI: 10.1147/rd.176.0525 38

[24] C. Bennett. Notes on Landauer's principle, reversible computation and Maxwell's demon. *Studies in the History and Philosophy of Modern Physics*, 34(3), 501–510, 2003. DOI: 10.1016/S1355-2198(03)00039-X 38

[25] P. Bergman and J. Lebowitz. New approach to non–equilibrium processes. *Physical Review* 99(2), 578–587, 1955. DOI: 10.1103/PhysRev.99.578 16

[26] M. Berry. Regular and irregular motion. In S. Jorna, (ed.), *Topics in Nonlinear Dynamics*. American Institute of Physics, 16–120, 1978. 15

[27] J.M. Blatt. An alternative approach to the ergodic problem. *Progress in Theoretical Physics* 22(6), 745-756, 1959. DOI: 10.1143/PTP.22.745 16

[28] R. Blatt and D. Wineland. Entangled states of trapped atomic ions. *Nature* 453, 1008–1015, 2008. DOI: 10.1038/nature07125 1, 32

[29] L. Boltzmann. *Lectures on Gas Theory*, 1895. Translated by Steven G. Brush. University of California Press, p. 60, 1964. 15

[30] F. Bonetto et al. Fourier's law: a challenge to theorists. In A. Fokas et al. (eds.) *Mathematical Physics 2000*. Imperial College Press, pp. 128–150, 2000. DOI: 10.1142/9781848160224_0008 14

[31] E. Borel. *Le Hasard*. Alcan, 1914. 15

[32] H.P Breuer and F. Petruccione. Concepts and methods in the theory of open quantum systems. *Lecture Notes in Physics*, 622, 65–79, 2003. DOI: 10.1007/3-540-44874-8_4 55

[33] H. Brown and J. Uffink. The origins of time–asymmetry in thermodynamics, the minus first law. *Studies In History and Philosophy of Modern Physics* 32(4), 525–538, 2001. DOI: 10.1016/S1355-2198(01)00021-1 13

[34] S.G. Brush. *The Kind of Motion We Call Heat*, volumes 1, 2. North Holland Publishing Company, pp. 366–377, 1976. 15

[35] J. Bub. *Interpreting the Quantum World*. Cambridge University Press, 1997. 16, 40

[36] J. Bub. Why the quantum?. *Studies in the History and Philosophy of Modern Physics*, 35, 241–266, 2004. DOI: 10.1016/j.shpsb.2003.12.002 40

[37] J. Bub. Quantum theory is about quantum information, *Foundations of Physics*, 35(4), 541–560, 2005. DOI: 10.1007/s10701-004-2010-x 39, 42

[38] S.H. Burbury. Boltzmann's minimum function. *Nature* 51, 320, 1895. DOI: 10.1038/051320a0 15

[39] O. Caldeira and A. Leggett. Quantum tunnelling in a dissipative system. *Annals of Physics*, 149, 374–456, 1983. DOI: 10.1016/0003-4916(83)90202-6 48, 49

[40] R. Clifton, J. Bub, and H. Halvorson. Characterizing quantum theory in terms of information–theoretic constraints, *Foundations of Physics*, 33(11), 1561–1591, 2003. DOI: 10.1023/A:1026056716397 35, 42, 47

[41] B.E. Davies. Markovian master equations. *Communication in Mathematical Physics*, 39, 91–110, 1974. DOI: 10.1007/BF01608389 12

[42] B.E. Davies. A model of heat conduction. *Journal of Statistical Physics* 18, 161–170, 1978. DOI: 10.1007/BF01014307 14

[43] D. Dieks. Communication by electron–paramagnetic–resonance devices. *Physics Letters* A, 92, 271. DOI: 10.1016/0375-9601(82)90084-6 6

[44] L. Diosi. Models for universal reduction of macroscopic quantum fluctuations. *Physical Review* A, 40, 1165–1174, 1989. DOI: 10.1103/PhysRevA.40.1165 48

[45] M. Dyakonov. Is fault–tolerant quantum computation really possible?. http://arxiv.org/abs/quant-ph/0610117, 2006. ix

[46] F. Dyson. Leaping Into the grand unknown. *The New York Time Review of Books*, 56(6), 2009. 20

[47] J. Earman. The problem of irreversibility. *Philosophy of Science (Supp)*, 2, 226–233, 1986. 16

[48] J. Earman and M. Redei. Why ergodic theory does not explain the success of equilibrium statistical mechanics. *British Journal of Philosophy of Science*, 47, 63–78, 1996. DOI: 10.1093/bjps/47.1.63 16

[49] J. Earman and J. Norton. Forever is a day, supertasks in Pitowsky and Malament–Hogarth spacetimes. *Philosophy of Science*, 60, 22–42, 1993. DOI: 10.1086/289716 52

[50] J. Earman and J. Norton. Exorcist XIV, The wrath of Maxwell's demon. Part I. *Studies in the History and Philosophy of Modern Physics*, 29(4), 435–471, 1998. DOI: 10.1016/S1355-2198(98)00023-9 37

[51] J. Edmonds. Paths, trees, and flowers. *Canadian Journal of Mathematics*, 17, 449–467, 1965. 52

[52] J. Emerson et al. Symmetrized characterization of noisy quantum processes. *Science*, 317, 1893, 2007. DOI: 10.1126/science.1145699 20

[53] D.J. Evans, E.G.D. Cohen and G.P. Morriss. Probability of second law violations in shearing steady states. *Physical Review Letters*, 71, 2401–2404, 1993. DOI: 10.1103/PhysRevLett.71.2401 26

[54] D.J. Evans and D.J Searles. The fluctuation theorem. *Advances in Physics*, 51, 1529–1585, 2002. DOI: 10.1080/00018730210155133 26, 27

[55] J. Fodor. Special sciences. *Synthese* 2, 97-115, 1974. DOI: 10.1007/BF00485230 52

[56] J. Fodor and Z. Pylyshyn. Connectionism and cognitive architecture: a critical analysis. *Cognition*, 28, 3–71, 1988. DOI: 10.1016/0010-0277(88)90031-5 52

[57] J.R. Friedman et al. Quantum superposition of distinct macroscopic states, *Nature*, 406, 43–46, 2000. DOI: 10.1038/35017505 46

[58] R. Frigg. A Field guide to recent work on the foundations of statistical mechanics. In D. Rickles (ed.), *The Ashgate Companion to Contemporary Philosophy of Physics*. Ashgate, pp. 99–196, 2008. 26

[59] C. Fuchs (2002). Quantum mechanics as quantum information (and a little more), http://xxx.lanl.gov/abs/quant-ph/0205039. 35, 39

[60] C. Fuchs and A. Peres. Quantum theory needs no interpretation, *Physics Today*, 53, 70–71, 2000. DOI: 10.1063/1.883004 35, 40

[61] F. Gaitan. *Quantum Error Correction and Fault Tolerant Quantum Computing*, CRC Press, 2008. 6, 10

[62] G. Gallavotti and E.G.D Cohen. Dynamical ensembles and stationary states. *Journal of Statistical Physics*, 80, 931–970, 1996. DOI: 10.1007/BF02179860 14

[63] R. Gandy. Church's thesis and principles for mechanisms. In J. Barwise et al. (eds.) *The Kleene Symposium*, North-Holland, pp. 123–148, 1980. 52

[64] J. Gea–Banacloche. Minimum energy requirements for quantum computation. *Physical Review Letters*, 89, 217901, 2002. DOI: 10.1103/PhysRevLett.89.217901 11, 30

[65] J. Gemmer, J. et al. *Quantum Thermodynamics*. Lecture Notes in physics 657, Springer, 2004. 14

[66] O. Goldreich. On quantum computing. Preprint, 2005. ix, 51

[67] S. Goldstein et al. Canonical typicality. *Physical Review Letters*, 96, 050403, 2006. DOI: 10.1103/PhysRevLett.96.050403 14

[68] D. Gottesman. Theory of fault–tolerant quantum computation. *Physical Review*, A 57, 127, 1997. DOI: 10.1103/PhysRevA.57.127 8, 18

[69] D. Gross, S. Flammia, and J. Eisert. Most quantum states are too entangled to be useful as computational resources. http://arxiv.org/abs/0810.4331, 2008. DOI: 10.1103/PhysRevLett.102.190501 27

[70] G.C. Ghirardi, A. Rimini, and T. Weber. Implications of the Bohm–Aharonov Hypothesis. *Nuevo Cimento*, 31B, 177, 1976. DOI: 10.1007/BF02728147 43

[71] G.C. Ghirardi, A. Rimini, and T. Weber. Unified dynamics for microscopic and macroscopic systems. *Physical Review* D, 34, 470–479, 1986. DOI: 10.1103/PhysRevD.34.470 43

[72] A. Hagar. A philosopher looks at quantum information theory. *Philosophy of Science*, 70, 752–775, 2003. DOI: 10.1086/378863 40

[73] A. Hagar. The foundations of statistical mechanics—questions and answers. *Philosophy of Science*, 72, 468–478, 2005. DOI: 10.1086/498474 14, 16

[74] A. Hagar. Experimental metaphysics$_2$. *Studies in the History and the Philosophy of Modern Physics*, 38(4), 906–919, 2007. DOI: 10.1016/j.shpsb.2007.04.002 xi, 39

[75] A. Hagar. Active quantum error correction: the curse of the open system. *Philosophy of Scinece*, 76(4), 506–535, 2009. DOI: 10.1086/648600 xi

[76] A. Hagar. To balance a pencil on its tip. Unpublished Manuscript, 2010. xi

[77] A. Hagar and M. Hemmo. Explaining the unobserved—why quantum mechanics ain't about information. *Foundations of Physics*, 36, 1295–1324, 2006. DOI: 10.1007/s10701-006-9065-9 45

[78] A. Hagar and A. Korolev. Quantum hypercomputation: hype or computation?. *Philosophy of Science*, 74(3), 347–363, 2007. DOI: 10.1086/521969 29, 52

[79] E. Hahn. Spin echoes. *Physical Review*, 80, 580–594, 1950. DOI: 10.1103/PhysRev.80.580 16, 18

[80] S. Haroche and J.M. Reimond. Quantum computing, dream or nightmare?. *Physics Today*, 8, 51–52, 1996. DOI: 10.1063/1.881512 5

[81] M. Hemmo and O. Shenker. Quantum decoherence and the approach to equilibrium, *Philosophy of Science*, 70, 330–358, 2003. DOI: 10.1086/375471 42

[82] M. Hemmo and O. Shenker. Quantum decoherence and the approach to equilibrium (II), *Studies in History and Philosophy of Modern Physics*, 32, 626–648, 2005. DOI: 10.1016/j.shpsb.2005.04.005 42

[83] A.P. Hines and P. Stamp. Decoherence in quantum walks and quantum computers. http://arxiv.org/abs/0711.1555, 2007. DOI: 10.1139/P08-016 20

[84] M. Hogarth. Non-Turing computers and non-Turing computability. PSA 94(1), 126–138, 1994. 52

[85] G. Kalai. Thoughts on noise and quantum computing. http://arxiv.org/abs/quant-ph/0508095, 2005. 20

[86] G. Kalai. How quantum computers can fail. http://arxiv.org/abs/quant-ph/0607021, 2006. 20

[87] G. Kalai. Detrimental decoherence. http://arxiv.org/abs/0806.2443, 2008. 20

[88] G. Kalai. Quantum computers, noise propagation and adversarial noise models. http://arxiv.org/abs/0904.3265, 2009. 20

[89] P. Kaye, R. Laflamme, and M. Mosca. *An Introduction to Quantum Computing*. Oxford University Press, 2007. 1, 9, 10

[90] D. Kastler. Foundations of equilibrium quantum statistical mechanics. In *Mathematical Problems in Theoretical Physics*, Springer, pp. 106–123, 1978. DOI: 10.1007/3-540-08853-9_9 12, 56

[91] J. Kempe. Approaches to quantum error correction. *Séminaire Poincaré*, 2, 1–29, 2005. 6, 8, 9, 24

[92] H. Khoon and J. Preskill. Fault–tolerant quantum computation versus gaussian noise. http://arxiv.org/abs/0810.4953, 2008. DOI: 10.1103/PhysRevA.79.032318 19, 20

[93] E. Knill. Quantum computing with realistically noisy devices. *Nature*, 434, 39–44, 2005. DOI: 10.1038/nature03350 11, 18

[94] E. Knill. On protected realizations of quantum information. http://arxiv.org/abs/quant-ph/0603252, 2006. DOI: 10.1103/PhysRevA.74.042301 30

[95] E. Knill and R. Laflamme. Concatenated quantum codes. http://arxiv.org/abs/quant-ph/9608012, 1996. 1, 9

[96] E. Knill, R. Laflamme, and W. Zurek. Threshold accuracy for quantum computation. http://arxiv.org/abs/quant-ph/9610011, 1996. 1

[97] E. Knill, R. Laflamme, and W. Zurek. Resilient quantum computation, error–models and thresholds. *Science*, 279, 342–345, 1998. DOI: 10.1126/science.279.5349.342 1, 9

[98] E. Knill, R. Laflamme, and L. Viola. Theory of quantum error correction for general noise. *Physical Review Letters*, 84, 2525–2528, 2000. DOI: 10.1103/PhysRevLett.84.2525 24

[99] E. Knill, R. Laflamme, A. Ashikhmin, H.N. Barnum, L. Viola, and W. Zurek. Introduction to quantum error correction. in N. Grant Cooper (ed.) *Information, Science, and Technology in a Quantum World*. Los Alamos, LANL, pp.188–225, 2002. 57

[100] A. Kolmogorov. Three approaches to the quantitative definition of information. *Problems of Information Transmission*, 1(1), 1–7, 1965. DOI: 10.1080/00207166808803030 17

[101] R. Laflamme et al. Perfect quantum error correction code. *Physical Review Letters*, 77, 198, 1996. DOI: 10.1103/PhysRevLett.77.198 8

[102] R. Landauer. Is quantum mechanics useful?. *Philosophical Transactions of the Royal Society, Physical and Engineering Sciences*, 1703 (353), 367–376, 1995. DOI: 10.1098/rsta.1995.0106 1, 5

[103] O.E. Lanford. The hard sphere gas in the Boltzmann–Grad limit. *Physica*, A, 106, 70–76, 1981. DOI: 10.1016/0378-4371(81)90207-7 14, 26

[104] J. Lebowitz. Exact results in non–equilibrium statistical mechanics: where do we stand?. *Progress in Theoretical Physics (Suppl)*, 64, 35–49, 1978. DOI: 10.1143/PTPS.64.35 14

[105] H.S. Leff and A.F. Rex (eds.). *Maxwell's Demon 2*, IOP, 2003. 37

[106] A.J. Leggett. Macroscopic quantum systems and the quantum theory of measurement, *Progress of Theoretical Physics Supp.*, 69, 80–100, 1980. DOI: 10.1143/PTPS.69.80 46

[107] A.J. Leggett. Macroscopic realism: what is it, and what do we know about It from experiment?. In R. Healey and G. Hellman (Eds.) *Quantum Measurement, Beyond Paradox*. University of Minnesota Press, pp. 1–22, 1998. 46, 48

[108] A.J. Leggett. Testing the limits of quantum mechanics: motivation, state of play, prospects, *Journal of Physics, Condensed Matter*, 14, R415–R451, 1998. DOI: 10.1088/0953-8984/14/15/201 45, 48

[109] A.K. Lenstra and H. W. Jr. Lenstra. *Handbook of Theoretical Computer Science, Volume A, Algorithms and Complexity* pp. 673–715, 1990. 5

[110] L. Levin. The tale of one–way functions. *Problems in Information Transmission*, 39, 92–103, 2003. DOI: 10.1023/A:1023634616182 ix, 17, 51

[111] D.A. Lidar, I.L. Chuang and K.B. Whaley. Decoherence free subspaces for quantum computation. *Physical Review Letters*, 81, 2594–2597, 1998. DOI: 10.1103/PhysRevLett.81.2594 23

[112] G. Lindblad. On the generators of quantum dynamical semigroups. *Communications in Mathematical Physics*, 48, 119–130, 1976. DOI: 10.1007/BF01608499 56

[113] S. Llyod. Ultimate physical limits to computation. *Nature*, 406, 1047–1054, 2000. DOI: 10.1038/35023282 30

[114] F.J. MacWilliams and N.J.A. Sloane. *The Theory of Error–Correcting Codes*. North Holland, 1977. 6, 9

[115] N. Margolus and L. Levitin. The maximum speed of dynamical evolution. *Physica*, D, 120, 188–195, 1998. DOI: 10.1016/S0167-2789(98)00054-2 11

[116] N.D. Mermin. Copenhagen computation: how I learned to stop worrying and love Bohr. *IBM Journal of Research and Development*, 48(1), 53–62, 2004. DOI: 10.1147/rd.481.0053 36, 40

[117] N.D. Mermin. *Quantum Computer Science, An Introduction.* Cambridge University Press, 2007. 6

[118] B. Misra and E.C.G. Sudarshan. The Zeno's paradox in quantum theory. *Journal of Mathematical Physics*, 18, 756–763, 1977. DOI: 10.1063/1.523304 24

[119] M. Nielsen and I. Chuang. *Quantum Computation and Quantum Information.* Cambridge University Press, 2000. 6, 18, 37, 47

[120] J. Norton. Eaters of the lotus: Landauer's principle and the return of Maxwell's demon, *Studies In History and Philosophy of Modern Physics*, 36, 375–411, 2005. DOI: 10.1016/j.shpsb.2004.12.002 37

[121] W. Pauli (ed.). *Neils Bohr and the Development of Physics*, McGraw Hill Book Inc. 1955. 36

[122] P. Pearle. Combining stochastic dynamical state–vector rreduction with spontaneous localization', *Physical Review*, A, 39, 2277–2289, 1989. DOI: 10.1103/PhysRevA.39.2277 43

[123] P. Pearle. True and false collapse, http://xxx.lanl.gov/abs/quant-ph/9805049. 42

[124] P. Pearle. How stands collapse I, II, http://xxx.lanl.gov/abs/quant-ph/0611211, 0611212, 2006. DOI: 10.1088/1751-8113/40/12/S18 43

[125] A. Peres, A. What's wrong with these observables?. *Foundations of Physics*, 33, 1543–1547, 2003. DOI: 10.1023/A:1026000614638 40

[126] I. Pitowsky. The physical Church thesis and physical computational complexity. *Iyyun*, 39, 81–99, 1990. 52

[127] I. Pitowsky. Laplace's demon consults an oracle: the computational complexity of predictions. *Studies in the History and Philosophy of Modern Physics*, 27, 161–180, 1996. DOI: 10.1016/1355-2198(96)85115-X 52

[128] I. Pitowsky. Quantum speed–up of computations. *Philosophy of Science*, 69 (Symposium), S168–S177, 2002. DOI: 10.1086/341843 29

[129] I. Pitowsky. Macroscopic objects in quantum mechanics: A combinatorial approach. *Physical Review*, A, 70, 022103, 2004. DOI: 10.1103/PhysRevA.70.022103 27, 41

[130] I. Pitowsky. From logic to physics: how the meaning of computation changed over time. *Lecture Notes in Computer Science*, 4497, 621–631, 2007. DOI: 10.1007/978-3-540-73001-9_64 52

[131] I. Pitowsky and O. Shagrir. Physical hypercomputation and the Church–Turing thesis. *Minds and Machines*, 13, 87–101, 2003. DOI: 10.1023/A:1021365222692 52

[132] S. Popescu et al. Entanglement and the foundations of statistical mechanics. *Nature Physics*, 2, 754–758, 2006. DOI: 10.1038/nphys444 14

[133] J. Preskill. Reliable quantum computers. *Proceedings of the Royal Society of London*, A, 454, 385–410, 1998. DOI: 10.1098/rspa.1998.0167 8

[134] J. Preskill. Fault–tolerant quantum computation against realistic noise. Talk given at the First International Conference on Quantum Error Correction, USC, December 2007. Available at http://qserver.usc.edu/qec07/program.html. 19

[135] H. Price. Burbury's last case: the mystery of the entropic arrow. In C. Callender (ed.) *Time, Reality and Experience*. Cambridge University Press, pp. 19–56, 2003. 14

[136] Z. Pylyshyn. *Computation and cognition: toward a foundation for cognitive science*. MIT Press, 1984. 52

[137] K. Ridderboss and M. Redhead. The Spin echo experiments and the second law of thermodynamics. *Foundations of Physics*, 28(8), 1237–1270, 1998. DOI: 10.1023/A:1018870725369 16

[138] R. Rivest, A. Shamir and L. Adleman. A method for obtaining digital signatures and public–key cryptosystems. *Communications of the ACM*, 21(2), 120–126, 1978. DOI: 10.1145/359340.359342 17

[139] G. Santayana. *The Life of Reason, Or, The Phases of Human Progress. Introduction and Reason in Common Sense*. Scribner's, 1905. 13

[140] E. Schrödinger. Probability relations between separated systems, *Proceedings of the Cambridge Philosophical Society*, 32, 446–452, 1936. DOI: 10.1017/S0305004100019137 43

[141] O. Shagrir. Computations by humans and machines. *Minds and Machines*, 12, 221–240, 2002. DOI: 10.1023/A:1015694932257 51

[142] O. Shenker. Interventionism in statistical mechanics—some philosophical remarks. University of Pittsburgh PhilSci Archive, http://philsci-archive.pitt.edu/archive/00000151, 2000. 15

[143] P. Shor. Algorithms for quantum computation: discrete log and factoring. *Proceedings of the 35th IEEE Symposium on Foundations of Computer Science*, 124–134, 1994. DOI: 10.1109/SFCS.1994.365700 5, 52

[144] P. Shor. Scheme for reducing decoherence in quantum Computer Memory. *Physical Review*, A, 52, 2493–2496, 1995. DOI: 10.1103/PhysRevA.52.R2493 1, 6

[145] P. Shor. Fault–tolerant quantum computation. *Proceedings of the 37th IEEE Symposium on Foundations of Computer Science*, 56–65, 1996. 8

[146] W. Sieg and J. Byrnes. An abstract model for parallel computations. *The Monist*, 82, 150–164, 1999. 52

[147] D. Simon. On the power of quantum computation. *Proceedings of the 35th Symposium on Foundations of Computer Science*, 116–123, 1994. DOI: 10.1109/SFCS.1994.365701 5

[148] C. Simon, V. Buzek, and N. Gisin. No signaling condition and quantum dynamics, *Physical Review Letters*, 87, 170405–170408, 2001. DOI: 10.1103/PhysRevLett.87.170405 47

[149] A. Steane. Error correcting codes in quantum theory. *Physical Review Letters*, 77, 793, 1996. DOI: 10.1103/PhysRevLett.77.793 1, 6, 8

[150] H. Tasaki. From quantum dynamics to canonical distribution: general picture and rigorous example. *Physical Review Letters*, 80, 1373–1376, 1998. DOI: 10.1103/PhysRevLett.80.1373 14

[151] B. Terhal and G. Burkard. Fault–tolerant quantum computation for local non–Markovian noise. *Physical Review*, A, 71, 012336, 2005. DOI: 10.1103/PhysRevA.71.012336 13, 20

[152] C. Timpson. *Quantum information theory and the foundations of quantum mechanics.* http://xxx.lanl.gov/abs/quant-ph/0412063, 2004. 42, 43

[153] R. Tomulka. Collapse and relativity. http://xxx.arxiv.org/abs/quant-ph/0602208, 2006. 47

[154] A.M. Turing. On computable numbers with an application to the Entschei-dungsproblem. *Proceedings of the London Mathematical Society*, 45(2), 115–154, 1936. DOI: 10.1112/plms/s2-42.1.230 51

[155] J. Uffink. Bluff your way in the second law of thermodynamics. *Studies In History and Philosophy of Modern Physics*, 32, 305–394, 2001. DOI: 10.1016/S1355-2198(01)00016-8 16

[156] J. Uffink. Compendium of the foundations of classical statistical physics. http://philsci-archive.pitt.edu/archive/00002691/, 2006. 25, 26

[157] W. Unruh. Maintaining coherence in quantum computers. *Physical Review*, A 51, 992–997, 1995. DOI: 10.1103/PhysRevA.51.992 1, 5

[158] L. Viola, E. Knill, and S. Lloyd. Dynamical decoupling of open quantum systems. *Physical Review Letters*, 82, 2417–2421, 2000. DOI: 10.1103/PhysRevLett.82.2417 23, 25

[159] L. Viola, E.M. Fortunato, M.A. Pravia, E. Knill, R. Laflamme, and D.G. Cory. Experimental realization of noiseless subsystems for quantum information processing. *Science*, 293, 2059–2062, 2001. DOI: 10.1126/science.1064460 32

[160] J. von Neumann. Probabilistic logics and the synthesis of reliable organisms from unreliable components. In W.R. Ashby, J. McCarthy and C E. Shannon (eds.) *Automata Studies*. Princeton University Press, pp. 43–98, 1956. 5, 19

[161] M.M. Wolf. Divide, perturb and conquer. *Nature Physics*, 4, 834–835, 2008. DOI: 10.1038/nphys1124 31

[162] S. Wolfram. *A New Kind of Science*. Wolfram Media, 2001. ix, 51

[163] W. Wootters and W. Zurek. A single quantum cannot be cloned. *Nature*, 299, 802, 1982. DOI: 10.1038/299802a0 6

[164] P. Zanardi and M. Rasetti. Noiseless quantum codes. *Physical Review Letters*, 79, 3306, 1997. DOI: 10.1103/PhysRevLett.79.3306 18, 23

[165] A. Zeilinger. The message of the quantum, *Nature*, 438, 743, 2005. DOI: 10.1038/438743a 39

[166] W.H. Zurek. Decoherence and the transition from quantum to classical, *Physics Today*, 44, 36–44, 1991. DOI: 10.1063/1.881293 37

Author's Biography

AMIT HAGAR

Amit Hagar is a faculty member in the department of History and Philosophy of Science (HPS) at Indiana University, Bloomington (IN). Before coming to Bloomington he taught at the University of Delaware and was a research fellow at the University of Konstanz. He was born and educated in Israel, where he received a B.A. and M.A. in philosophy from the Hebrew University of Jerusalem. His Ph.D. thesis, entitled "Chance and Time" (2004), was written in Vancouver in the University of British Columbia and concerns the foundations of statistical physics. His main interests span the foundations of statistical and quantum mechanics, the philosophy of time, and the notion of physical computation, especially in the context of quantum information theory and quantum computing.